T0088005

SAFER
SKIES

An Accident Investigator on Why Planes Crash and the State of Aviation Safety

David Soucie
WITH
Ozzie Cheek

SKYHORSE PUBLISHING

Previously published by Skyhorse Publishing as *Why Planes Crash* 2011

www.skyhorsepublishing.com

10 9 8 7 6 5 4 3 2

Library of Congress Cataloging-in-Publication Data is available on file.

Print ISBN: 978-1-63220-589-6
Ebook ISBN: 978-1-63450-016-6

Printed in the United States of America

NOTE: The names and other identifying details of some people and companies in this book have been fictionalized. All fictionalized names are marked with an asterisk at the first reference. Some people have asked that I not reveal their names. Sometimes the character is minor, and I've forgotten the real name. Some names have been changed to protect the safety and well-being of my family. This book is a memoir of my life, but some sequences and events have been condensed or changed to fit the constraints of the editing process.

Childhood drawing by my son, Tyler

I dedicate this book to my lovely bride, who truly knows my heart; and to my loving son, who truly knows my spirit.

Contents

Introduction

Do you fear flying? You are not alone.

According to the National Institute of Mental Health, 6.5% of Americans have a fear of flying so severe that it qualifies as a phobia or anxiety disorder. A July 2007 article in *The New York Times* about the fear of flying noted that 40% of the people have some degree of anxiety about flying. The same article noted an international fear-of-flying-conference held in Montreal that year, organized by Dr. Lucas van Gerwen, an aviation psychologist and professional pilot in the Netherlands, reported lower-than-expected attendance. Why? Many did not want to fly to the fear-of-flying conference.

If the frequency of airline crashes is the basis for a fear of flying, 2014 should prove to have a great increase in strictly earthbound travelers. Yet, this does not appear that this will be the case.

Air travel in both the United States and Europe has increased in every category. We have short memories. Even the most tragic disasters, like the disappearance of Malaysia Flight 370 and the shooting down of Malaysia Flight MC17, quickly fade in our memories. Fortunately for those who fear flying, yet must fly to earn their living, the memories do fade. *It is selective memory that culls out the disturbing reality of disastrous loss of life.*

For safety regulators and airline managers and owners, the memories do not fade, and that is how it should be. They must be hypervigilant about using past accidents as lessons learned in order to do everything in their power to prevent future accidents. Since this book was originally published three years ago, there have been a vast number of safety improvements by airlines, including improvements in technology and safety procedures and systems and better communication and sharing of information.

According to Aviation Safety Network, an organization that tracks aviation statistics, 2012 was the safest year since 1946 for commercial aviation deaths worldwide. Then came 2013, and it was a safer year for commercial aviation deaths worldwide—the safest year since 1946. While United States' commercial airlines have continued to maintain exemplary safety records, 1,253 lives have been lost to tragic aircraft accidents worldwide thus far in 2014. News of these tragic accidents brings back those faded memories, so those who fear flying—aviophobics—have additional fodder to stoke their fears.

Those who are concerned with airline safety can never rest. I will forever ask myself what can be done to make the skies safer. In the business of airlines safety, you never know what you did right—you only know when you did something wrong.

CHAPTER ONE

Bird on a Wire

From across the room, I could see Jill's smile through the Christmas tree tinsel. She was reading *A Christmas Carol* to our son, Tyler. Her eyes met mine, and her smile disappeared. After five years of marriage, she knew when something was wrong. "David," she said, "are you okay?"

I dropped the telephone receiver on the floor and fell to my knees. Tears filled my eyes.

Jill tossed the book aside and rose from her chair near the fireplace. "You're scaring me, sweetie," she said. "What is it?" I looked at her, but I couldn't speak. She hurried over and knelt beside me. "Breathe!" she yelled. "You're not breathing, David!"

I gasped for air. "They went down. Mike and them. Mike Myers hit a wire and now . . . oh god!"

"Who else?" Jill asked desperately, while hugging me. "Who was with him?" Jill knew Mike and his family, as well as the flight nurses and the crew in Grand Junction, Colorado.

"I don't know. Some patients, I think. A nurse. They may all be dead." That's when the truth smacked me full force. Mike had crashed after hitting a power line or telephone wires because the helicopter was not equipped with a wire-strike kit, a safety addition that I had refused to add. "I think it's my fault," I said, feeling numb. "It's my fault they went down."

"Don't be silly," she said. I heard a faint voice coming from the phone on the floor, but I couldn't bring myself to touch it. "Are you going to pick it up?" Jill asked. I shook my head no, so she picked up the receiver and listened to whatever else Roy Morgan was saying. "Okay, Roy," I heard her say. "We're leaving now."

My brain processed only the simplest parts. Leaving? Check! Meeting Roy? Check! Roy Morgan owned Air Methods Inc., the emergency medical helicopter company where I worked. I was twenty-six years old and already the senior director of maintenance. Mike Myers was one of our pilots. He flew helicopters that I was responsible for maintaining.

While I bounced between despair and disbelief, Jill bustled around. We had a child to care for. We had to get to Arapahoe County airport to meet Roy Morgan so we could fly to Grand Junction right away. I had to go see Mike Myers. I had to fix things. Fix things? Check!

Jill grabbed coats and car keys. "Dave," she said from the door, "you ready?" I realized at that moment that whatever I would do in life that was worthwhile would be because of Jill's steady presence. I nodded, looked one last time at the glowing tree, and followed her out.

It was Christmas Eve 1985.

The December weather was bone-chilling cold in the West. An ice storm had turned the highways into asphalt skating rinks, and while weather conditions made night flying more dangerous than usual, flying was faster and safer than using the highways. Besides, pilot Mike Myers was highly experienced on the Bell 206 helicopter. That's why he was transporting two young flight nurses, Debbie Carrington★ and Deana Smith★, to a remote region near Monticello, Utah, to assist a pregnant woman with a premature birth.

Nobody wanted to work the holidays, but in the business of emergency medical care, holidays were often a busy time. The best Mike could do was promise his family that he would be back home by Christmas morning. When he left his house, he had little reason to believe that he would not keep his promise.

Despite the weather, the night trip from St. Mary's Hospital in Grand Junction to Monticello was flawless. The nurses arrived in time to assist with a difficult delivery, and before long, the new mother and her baby were doing well enough that Debbie and Deana could leave. "I'll prep the helicopter while you two get packed up," Mike told them.

Dr. Johns*, the attending physician at the Utah clinic, walked with Mike out the back door to the helipad, where Mike checked the sky. The winter air in the West is dry and crisp, the sky often cloudless.

On this night, the sky was bright with a full moon, and reflections off the ice and snow made it all the brighter.

While Mike stood there, wind swooped down off a hill behind him. The clinic is located in a box canyon, so taking off in the helicopter meant contending with high terrain and power lines. The helipad was elevated, but given the weight he would carry, the location of the helipad, and the weather conditions, Mike wanted to be certain he could clear the power lines on takeoff. He headed down to the street below for a closer look.

The staircase was icy, so Mike carefully held on to the railing. He stopped at the bottom and kicked the snow from his boots. The wires overhead swayed and hummed. Icicles that dangled from the wires like shiny Christmas tree ornaments broke loose and plunged to the street below, shattering.

The two flight nurses had already settled into their seats by the time Mike gingerly made his way back up the steps. "Looks like everything's a go," he told Dr. Johns. They exchanged Christmas greetings, and then Mike climbed into the pilot's seat, strapped on his seat belt, and started the engine. The time was 11:30 PM on Christmas Eve.

Dr. Johns had used the helicopter service before, so he was familiar with the helicopter's sound. The metal blades against the dense, cold air sounded like a bullwhip cracking. Usually, the sound lessened as the helicopter rose into the air. On this night, the blades continued to crack and pop loudly as the helicopter failed to climb.

Liftoff is influenced by the weight of the air, which is influenced by temperature, humidity, and other factors. To make matters worse, under certain conditions, ice can coat the helicopter blades, rendering them useless. Mike strained to pull up on the collective control to gain altitude, but instead, the helicopter slowly began to descend. The ice-laden power lines were directly below.

Between the noise from the blades and the buzzing wires, any screams from inside the helicopter never reached Dr. Johns. Still, he watched from outside the clinic door, horrified, as the power lines snared the helicopter. The impact launched the tail of the chopper high above them. Sparks and electrical fires lit up the sky. The pole that Mike had earlier used to dislodge the snow from his boots whipped violently under the impact. For a second, the helicopter perched on the wire like a tired bird. Then the wires snapped. The noise was louder than gunshots.

The bubble of the helicopter is Plexiglas. It is durable and strong, yet a snapped wire cut through the plastic windshield as if it were made of hot wax. The wire cut not only through the Plexiglas and the instrument panel, but also deeply into pilot Mike Myers. The helicopter fell helplessly to the street below.

An emergency call went out. Eventually, a helicopter from St. Anthony Hospital in Denver arrived to transport Mike Myers, Debbie Carrington, and Deana Smith back to St. Mary's Hospital in Grand Junction.

This time, they were making the ride as patients. When they arrived in Grand Junction, all three were admitted to the hospital. Within a short time, Debbie Carrington's condition was upgraded from critical to serious, while Deana Smith's was upgraded from serious to fair. The injuries to Mike Myers, however, could well prove fatal.

Roy Morgan and I arrived at St. Mary's Hospital shortly after 10:00 AM on Christmas Day. I stood outside the trauma room and watched Roy hold Mike's hand as he talked to him. It didn't appear that Mike was listening. Instead, his eyes were looking through the glass door directly at me. Roy turned and eyed me curiously and then motioned for me to come in. I was hesitant to enter the room. I had not admitted to my boss what I feared. I had not told him that I was responsible.

Two months earlier, I had been sitting in a meeting listening to Mike Myers. "Dave," he said, "we need to install wire-strike kits." The wire-strike kit is like giant scissors that fits on the front of the helicopter to cut through electrical and communication wires and cables before they can snare the rotor or body of the helicopter and bring it crashing down. I knew why he wanted it. I also knew how much it would cost to install the kits on the three Bell 206 LongRanger helicopters that Air Methods owned and operated.

In addition to Mike Myers, several other pilots attended this meeting. They were mostly veteran pilots adorned with handlebar mustaches and wearing faded green jumpsuits. Unlike them, I was young and had little experience. Still, I tried to sell them on my decision.

The meeting was held in October in our Aurora offices. The gold-leafed aspens were already bare, but winter had not yet gripped Colorado. The Christmas tragedy was still to come, and Mike Myers was not yet lying in a hospital bed, looking at me with knowing eyes. "Buying wire-strike kits is not economically justified," I told Mike that day. "If the expenditure isn't already in my budget, it has to be justified."

Air Methods was a small but growing company at the time, and as the director of maintenance, I had to decide if wire-strike kits were more important than other demands on the budget. My decision was not callous or cavalier. It was based on sound business practices. Air Methods purchased only new helicopters and only from Bell. Our purchase agreements included both maintenance and warranty. Neither Roy nor I had budgeted for additional equipment. "Wire-strike kits aren't even part of the optional equipment when you buy a new helicopter," I said.

The expressions on the faces of the pilots told me that my answer wasn't what they wanted to hear, so I took another approach.

"We've never had a wire strike or even come close to one, have we?" There was silence in the room. "If you can't demonstrate a real risk, how can I justify spending the money?" Clearly, my argument was rational, but it was also falling on deaf ears. *Okay*, I thought, *if the high road is blocked, take the low road.* In aviation, the low road means you blame the Federal Aviation Administration (FAA). "These kits are not required by the FAA."

Even if the pilots felt it was wrong not to buy the kits, everyone there knew that the FAA was the law in aviation; and if the law didn't require us to do something, we usually didn't do it. I had learned early on that the FAA was always a good scapegoat. The truth was that I had no idea how to weigh the value of an investment in safety equipment.

All I knew was that safety equipment cost money and didn't provide a measurable return on investment.

At this point, Mike Myers and two other pilots, Walt Wise and Steve Scully, left the meeting in disgust. I continued to rant even as they were leaving. "They don't improve efficiency," I said. "They don't even look good." I didn't know enough at the time to shut up once I had won. "The pilots should just be more careful," I added. "All wires have those orange balls on them, don't they?" Nobody said anything, but there were grunts and groans as the remaining pilots got up to leave. "Well, don't they?"

The pilots left in dismay. They realized something that I didn't—they realized I was unable to see the blind spot in my own economic system and its logic. A blind spot is just that. You can't fix something you can't see.

I could have told them that Air Methods had serious money problems. I could have said that in the aviation industry, safety is regularly sacrificed for profit. In time, I would learn just how often the issue of safety versus profit arose during the investigation of aviation disasters. In the beginning, I really believed that being like everyone else in the aviation industry was something of which I could be proud. I believed that by following FAA guidelines for meeting the minimum standards, I was doing a fine job. I just didn't realize that part of my job was to estimate the monetary value of a human life. I didn't realize that until I walked into Mike Myers's hospital room.

Mike was strapped to the bed and hooked up to tubes. The drone of the machines gave me a reason to look away from him as I approached his bed. Mike took my hand. He was very weak, but he managed to squeeze my hand with surprising vigor until I looked at him. "Tell them, Dave. Tell them what happened." He labored to speak. "You tell my wife and kids I'm not coming home." I started to say something—I don't know what—but he squeezed my hand again to make sure I was listening. His eyes seemed to bore into me. "This could have been avoided, Dave," he said.

I nodded my head and looked over at Roy Morgan until I could manage to utter two words. "I know," I said to Mike. Then I left the room and stood in the hallway where Mike's eyes could not find me.

A little while later, Roy joined me. Roy could see that the thought of telling Mike's wife and three children that he wouldn't be home for Christmas, or ever again, was more than I could handle. "I'll do it, Dave," he said. "I'll talk to the family." I nodded again. I had used up all my words. I had also made a promise to Mike Myers that I couldn't keep. Not then. Not until now.

Since I was a boy, I had been having recurring dreams of flying among tall buildings, only to hit a wire that sent me spiraling to the ground like a fallen bird. In my dreams, I was the one falling. In real life, it was my friend Mike Myers who toppled off the wire and fell to earth. He died at 3:00 PM on Christmas Day.

Even then, I knew that Mike's admonition to tell "them" what happened entailed more than my accepting the blame for the accident. I did accept it. He knew that. I think he was also asking me to change the way decisions were made in the aviation business. He wanted to be sure I accepted the challenge as well as the blame.

This tragedy changed the course of my life and set me on a long journey of discovery. I became a passionate student of the complexities and interdependencies of hazard, probability, and risk. Over the following years, I was driven to learn more about how to recognize hidden accident indicators or precursors that could make business decision makers and regulators aware of a possible accident, to understand what those indicators tell them about imminent threats to safety, and to find ways to prevent an accident before it happens.

Improving aviation safety is my response to Mike Myers. It is my way of saying yes to Mike's challenge. My understanding of how to make the skies safer and the invention of a method of eliminating the root cause of most accidents were still some years away, but my battles with the FAA and the aviation industry began Christmas Day in a hospital room. They began the day that Mike Myers wasn't coming home.

CHAPTER TWO

Faster Than a Speeding Bullet

Louis Cushman, my uncle Louie, had a ranch near Longmont, Colorado. He was the real deal, a larger-than-life man with a weather-beaten face and a hitch in his walk from a bull-riding incident a few years back. He wore a sweat-stained cowboy hat that he removed only as part of the ritual for saying grace at the supper table. This ritual was straight out of the 1965 movie *Shenandoah*: "We thank you, Lord, for this food. We cleared the land. We tended the herd and planted the corn, cared for it, and harvested it. Then my wife prepared the meal, and we thank you for the opportunity to do all that. Amen."

"Amen," we would all reply, for Jimmy Stewart had nothing on Uncle Louie.

In 1970, when I was eleven years old, I spent many summer days at the ranch with my cousin Kathie, floating down the High Line Canal or bailing out of tire swings or talking beneath the huge oaks that shaded the lazy water.

One time, Uncle Louie discovered that we had played all day instead of doing the chores he had asked us to do. At the supper table, after removing his hat and saying grace, he told us that he was disappointed. That was all he said. Supper was quieter than usual.

After we finished eating, I gathered up the courage to ask him what he did for fun. It was a question only an eleven-year-old would ask a man like Uncle Louie.

"Fun?" he repeated. "Oh, I have plenty of fun." He cocked his head and gave me a look that said I didn't even know the meaning of the word. "I have more fun before breakfast than you've ever had."

Both his voice and his look scared me a little. It was as if he were saying, you asked for it, kid. "You get a good night's sleep, son," Uncle Louie told me, "and I'll pick you up at 5:00 AM sharp. We'll have us some fun."

I nodded in agreement, not knowing what else to do, and soon left with Aunt Monie. She delivered me to my home away from home, my grandparents' house. Grandpa and Grandma Conard lived in a small tract house next to the remains of a cornfield overrun by suburban sprawl.

Long after I settled into the sleeper on the back patio, Uncle Louie's promise kept me awake. I wondered what it was. What does he do that ignites that fire in his eye? I finally drifted off to sleep.

At 5:00 AM, Uncle Louie pounded on my grandparents' back patio door. "Good morning, sunshine!" He often teased me about my long blond hair and told me that I had been born with sunshine up my butt because I was always smiling and happy. That morning, I awoke neither smiling nor happy. I was sunburned and tired from the long weekend spent with Kathie on the canal and had forgotten why he was there. Uncle Louie had no patience for that. He wrapped me in the homemade patchwork quilt that Grandma Conard had put over me during the night, threw me over his shoulder, and packed me into his Jeep. We headed off down the road.

Nothing was said during the first half hour. Maybe I even dozed off. Maybe that's why I jumped when Uncle Louie barked, "Get dressed. We're almost there." I struggled to put on my pants and shoes as we bounded over a rutted trail. "You see that?" Uncle Louie asked. "See those lights? Hear it?" He stopped the Jeep, and there it was.

"It's an airplane!" I screamed.

"Yup." Moments later, Uncle Louie stood proudly beside the door of the small plane, his hands on his hips and his head held high. It was a pose that reminded me of Superman, one of my favorite TV characters. Fighting for truth and justice didn't mean much to me then, but "faster than a speeding bullet" and "able to leap tall buildings at a single bound"—well, sign me up!

We boarded the airplane, and I watched Uncle Louie finish a checklist and start the engine. He told me to push a big white knob forward. When I did, the engine roared and the airplane began to move. I tried to turn left with the control wheel, but the plane went right. I tried to turn right, but the plane went left. I looked over at Uncle Louie and yelled, "Help me!"

He laughed. "I was messing around, Dave. You steer with your feet. See?" He pointed to the floor, where I could see him pushing with his left foot to go left and his right foot to go right. Uncle Louie guided the plane to the runway and then held the microphone to his mouth, "November six-one-two tango whiskey in position."

A voice on the radio answered immediately. "November six-one-two tango whiskey cleared for takeoff."

"Push it now," Uncle Louie told me, indicating the knob. "All the way, son. Let's go flying." He kept the brakes on while I pushed the knob into the panel. The engine roared even louder, and when Uncle Louie let go of the brakes, I was slammed back in my seat. We rolled faster and faster down the runway until Uncle Louie pulled the control wheel to his chest and we rose into the air.

Uncle Louie adjusted some knobs on the floor and another one on the dashboard in front of me. He tapped lightly on the control wheel and said, "Yup, that'll do it." I studied his huge weathered and scarred hands on the wheel. He turned left and right with a touch so delicate that it seemed as if he were willing the plane to turn and having it obey. As he flew, I studied the dashboard's placards and warnings—Mixture, Throttle, Gear, Flaps. They made no sense to me. One placard above the control wheel mounting said, "To make houses smaller, pull back; to make houses larger, push forward." I giggled, and Uncle Louie looked over at me and smiled.

After flying about thirty minutes, Uncle Louie yawned and stretched both hands up in the air behind his head. Both hands? Before I could stop myself, I let out a yelp. "It's all yours, Dave," he said.

I grabbed the wheel and pulled it back. We went up. "Wow!" I yelled. "The houses really do get smaller."

Uncle Louie told me to take it easy and not to climb so fast, so I pushed the wheel forward, and we went into a violent dive. I took us up and down like this several times before Uncle Louie, turning a shade of green, took the wheel. "Very lightly," he said. "Like this." He demonstrated the touch and feel of flying by using his big fingertips. Then he let me try again. Before long, I could do it, willing the airplane slowly up or down, left or right. I was flying. I was Superman.

Despite love at first flight, I didn't fly again for many years. My father and my mother divorced in the summer of 1970, the summer of flight, and Mom was left to raise six kids on her own. With barely enough food on the table and all of us wearing hand-me-down clothes from our cousins, flying lessons were out of the question. Long summer stays with Uncle Louie were also out of the question. After the divorce, Mom needed me at home.

A few years later, when I could drive, I would stop at the Jefferson County Airport in Broomfield as often as possible to watch the airplanes. One day during my senior year of high school, instead of sitting in the parked car watching, I followed a small airplane as it taxied across the road toward Colorado Aero Tech. Colorado Aero Tech is a mechanic school with a flight club. To this day, I'm not sure what motivated me to follow the plane, but once I saw the building, I knew I had to get inside.

I told Mom about the school, and she soon took me there. It was everything I had imagined: real airplanes being disassembled, engines being run on test stands, and rivet guns banging aluminum against steel bucking bars.

A short time later, the admissions counselor signed me up for the classes and the flight club. He then asked Mom for a five-hundred-dollar deposit. Five hundred? I knew she didn't have it.

Mom dug through her purse. "Oh, I forgot my checkbook," she said. She dug around some more, opening several secret compartments, and emerged with a hundred-dollar bill. I had seen Grandpa Conard slip her the money during our last visit to his house in Longmont. The counselor said one hundred dollars would be fine. Mom's eyes never left that bill as he slowly opened his cash box, placed it in the drawer, and closed it. When it snapped shut, Mom shuddered as if she had just awakened from a dream. I felt so selfish when I realized she handed over all the money she had.

In my excitement, I had forgotten all about my family: about Clay, my older brother in the navy, who sent us his paycheck; about Theresa, who was trying to find the money for college and design classes; and about Gracie, who worked late shifts at a Sambo's restaurant to earn the money for her cheerleading skirts. I forgot about my young sisters, Gayle and Lynn, who didn't have a dad around. I forgot about everything except Colorado Aero Tech.

Over the next few weeks, I begged and borrowed money for school. I needed about $2,800. I managed to raise some money but not nearly enough. I had one last chance. Mom's sister, my aunt Gay, and her husband, Carl Morris, lived in Bartlesville, Oklahoma, where Uncle Carl was a successful executive for Phillips 66. He was the only relative I knew who had enough money to help me with school. I called him and told him my plan.

"Why do you want to be a pilot?" he asked.

"Ummm . . . because I like flying?" I wanted to say whatever Uncle Carl was waiting to hear.

My answer was not good enough. "I'll loan you the money to go to school and get the mechanic's license," he informed me, "but not for flying lessons." As an incentive, he went on to say that he "might" be willing to loan me the money for flight school after I completed my studies and obtained my mechanic's license. Although I was a bit surprised that he charged me 3 percent interest, it was a great offer, and I was happy to accept it.

First, I had to finish high school. When I learned that I could finish school early if I attended summer classes, I jumped at the opportunity. I finished my credits and was told I would graduate number 495 out of 540. I was ecstatic! Top 10 percent! It was actually the opposite—the bottom 10 percentile. But top or bottom, finishing high school early meant I could start school at Colorado Aero Tech in November 1976.

I was not just enrolled in classes, but I excelled in them. Even while working two jobs, I maintained a 3.9 grade point average at Colorado Aero Tech. Life was simple and busy. I went to school from seven to two thirty, hung downspouts from three to six, and then worked at my uncle Mike's Texaco station from six until ten. This was my schedule for over a year.

I was seventeen when I passed the Federal Aviation Administration (FAA) Airframe and Powerplant tests for certification as a mechanic. Although I scored 98 percent on the tests, I was denied a license. FAA regulations stipulate that you have to be eighteen or older to be licensed. This seemed unfair, and I confronted the FAA—for the first time.

The receptionist at the FAA General Aviation District Office (GADO) in Broomfield wore a blue-and-white nametag that said Mary*. "Mary, this rule just doesn't make sense," I argued. I was young, and I assumed that regulations were supposed to be sensible. "A pilot can get a license at fifteen," I said, my voice getting louder. "How come I can fly at fifteen but I can't work on a plane until I'm eighteen?"

"Because the pilot'll probably just kill himself." I looked around for the source of the words. They had come from a gruff-looking, older man who had just entered the office. He was slightly bent over, tall, and thin, and the smoke from the cigarette hanging precariously from his lips seemed to stick to the hair cream that plastered his sparse, gray hair to his head. He struggled to set down the two boxes he carried without interfering with the cigarette in his mouth. "What seems to be the problem, Mary?" he asked.

"This gentleman finished his test for the A&P, but he's not eighteen yet, so I can't issue the license."

The man told me that his name was Wayne McCannon and that he was an aviation safety inspector for the FAA. "You take care of these tox-boxes, and I'll talk with him," he told Mary. I had no idea what a tox-box was. In a few years, I would become intimately familiar with a tox-box.

McCannon leaned over the counter and put his face directly in front of mine. "Listen, kid, I'm going to let you in on a secret." His piercing eyes were frightening enough, but his smell was far worse. I didn't know it then, but it was the smell of death. "The FAA doesn't care what you or anyone else thinks about our rules. We're here for one reason and one reason only." I followed his dirty, bony finger pointing to the two strange boxes. "To make sure you don't end up in one of those boxes over there, like Mr. John Doe and his daughter." After puffing his cigarette, smoke rolled out of his nose and mouth as though he was a fire-breathing dragon. "Now, what's your problem?"

When I had walked in, I was a confident seventeen-year-old. Now I heard a voice as cracked as a prepubescent teenager's say, "Nothing. Never mind."

Until that day, I had thought of flying as fun, a chance to be free from the earth and to command the universe from high above. Wayne McCannon gave me my first glimpse of the pain and suffering that flying can cause. Although unknown to me at the time, this was the beginning of my lifelong passion—some would say obsession—for improving aviation safety. It was not my plan, but it was my fate.

When I turned eighteen, I received my A&P license and was ready to work on airplanes. The thought of being a pilot had vanished; I was an aviation mechanic. At least I would be if I had the tools. Aviation mechanics need expensive tools, and I had only a few hand-me-down wrenches and sockets. Although my father dropped out of our lives when I was twelve, my mother kept in touch with him, and she informed him that I had graduated from Colorado Aero Tech and that I needed tools and a toolbox.

My father was a kind man with a big heart, but over the years, alcohol had taken a heavy toll. The only thing we could count on was for Dad to make poor choices. When he appeared unannounced at my door not long after I had graduated, I was greatly surprised. "Come on out to my car, Dave," he said. I followed him to his car, where he opened the trunk and proudly displayed a pile of new tools. None of the tools was a name brand—no Snap-on or Craftsman—and there wasn't a full set of anything, just bits of this and that. Even so, I knew that he was offering me everything he had to give. Suddenly, all the anger I had built up toward him over the years dropped away. I strongly suspected that he had stolen the tools from the train yard where he worked, but I didn't ask. I just smiled as I filled a cardboard box with the tools. I took them inside and put the box in a closet. I never opened it again.

CHAPTER THREE

Flight-Test Dummy

In 1978, I got the most prestigious job a mechanic could have. I was working on Learjets at Combs-Gates in Denver. Although it was my first job after graduation, within three months, I was assigned a custom installation on the newly designed Learjet 35. The revamped design included a camera door that used an innovative style of vortex generator on the wings to prevent uncontrolled stalling in high-altitude flight. My task was to install the equipment.

I had just finished the installation and had stepped back to admire my work when a commanding voice from behind me said, "Good job!" The praise was coming from Les Jordan, a TOP GUN pilot and head salesman for Learjet. "Listen, kid," he said, "come go for a ride. I need some moveable ballast on board for the test flight." I jumped at the chance.

I towed the airplane outside and positioned it perfectly for the photographer to get a shot of my custom installation with the sunset in the background. By the time I boarded the plane, my head, inflated with self-importance, barely fit through the door.

Within minutes of takeoff, we had climbed so high ice was building up on the seat rails near the floor, and the crystallization on the small portal windows made them appear to be cracked. When we reached a frigid fifty-two thousand feet, the Learjet stalled. For a while, there were minor bumps and a slight twist here and there, but all in all, the stall test was not as bad as I had secretly feared. Just as I relaxed, Les told me to move. "Let's see how relocating you affects the stall characteristics."

I loosened my belt, got out of my seat, and was moving toward the front when the nose of the aircraft, pointed upward, reversed and slammed straight down. My feet flew up. I grabbed the strut holding the seat to the floor but found my legs above my head. I was totally weightless.

I looked out the front of the cockpit as we plummeted toward the earth. Seconds later, the blue waters of Horsetooth Reservoir, west of Fort Collins, Colorado, filled my window view. I saw the wake of a boat and the spray of a water skier sweeping from left to right behind it. I knew we were going to crash. I knew we were going to wipe out the boat and the water skier. I knew we were all going to die.

I tried to say something to Les, but before I found the words, the power levers slammed back, the lift spoilers deployed, and the controls were pulled into his chest. The wheel shook violently. I was thrown to the floor. Looking up, I still could see out the cockpit windshield as the plane leveled. I saw the boat give way to empty blue water and water to beach and beach to dirt and trees, and then I saw a mountain! A mountain is known as a "granite cloud"—the top pilot killer. We were headed straight toward it. My big head now felt as small as my shrunken testicles.

Les suddenly reversed everything he had done earlier: the spoilers came in, the throttles went forward, and I went backward against the rear bulkhead. The cockpit windshield was still filled with the mountain of which we were about to become part. I reached for the seat belt to prepare for impact, as if a seat belt would save me from a five-hundred-mile-per-hour kiss with a stone monolith.

At the last possible moment, Les pulled the control yoke toward him, and the mountain gave way to sky. Blue sky! Nothing but sky! I heard Les laughing. I was too pissed off to say anything until we were on the ground.

Once Les had parked the plane, I started to launch into a tirade about his antics, which was when he slapped me on the back and said, "Dave, there's no such thing as moveable ballast." Then I got it. The whole flight had been a big joke—on me. I was simply a flight-test dummy.

Les Jordan's ride was a lesson in humility, but I was still too young and arrogant to realize that humility is the most valuable tool in the mechanic's toolbox. All humans make mistakes; we make them constantly. Yet mistakes are rarely fatal when we're on the ground. Mistakes made in the air are less forgiving. A good mechanic must imagine the effect of mistakes at fifty thousand feet while going five hundred miles an hour in a pressurized metal tube filled with explosive fuel. My next lesson wasn't long in coming.

After I had been at Combs-Gates a while longer, I was assigned a detailed service inspection on a Learjet 25. One of the routine service items involved removing the outflow valve for cleaning. The outflow valve controls the pressure in the cabin. Standard atmospheric air pressure is 14.7 pounds per square inch. This means that every square inch of skin is experiencing this amount of pressure at all times. Since air pressure affects nearly everything in pressurized airplanes, including temperature and oxygen level, control of air pressure is critical. At eighteen thousand feet, natural atmospheric pressure is only 7.35 pounds per square inch, about the same as soccer ball pressure. We rely on the outflow value to correct this. We're humans, not soccer balls. Because the outflow valve can fail, just as any part can, Learjet also installed a safety valve on their plane. It had to be removed periodically and cleaned or it could stick. The cleaning and installation of the safety valve was one of my jobs.

I completed the detailed service inspection and notified Ridgerunner*, our chief inspector, that the plane was ready for a test flight. I am not sure why everyone called him Ridgerunner. I never asked, but I assumed it had something to do with the sharp ridge of his neatly combed crew cut. Ridgerunner and our pilot, Kevin Finn*, and I climbed into the Learjet 25. I sat in the backseat. All I could see from where I sat were Ridgerunner and Kevin's hands on the throttles.

The engines spooled, and my ears felt the pressure increasing. When the aircraft started rolling, the pressure increased dramatically. I felt dizzy and disoriented. I saw Ridgerunner's left hand—he was in the copilot seat—drop off the throttle. Then I saw Kevin's hand slide off the controls. Standard procedure is for both pilot and copilot to hold the controls until after takeoff. Something was terribly wrong.

I yelled at them to abort takeoff, but there was no response. I released my seat belt and crawled on my knees past two rows of empty passenger seats to the front. By this time, I could barely focus, but I could see that neither man was doing anything to control the plane. We were traveling at more than ninety miles per hour and drifting badly to the right side of an elevated runway that crossed directly over Interstate 70. In fact, we were headed straight toward the bridge. I reached the throttle controls and pulled them back. "Ridgerunner!" I yelled.

Ridgerunner awakened and realized what was happening. He knew that using the brakes would do little good since we were not straight on the runway. His only hope of avoiding an accident was to engage the thrust reversers. That's what he did. The clamshell doors closed with a loud bang as they came together behind each of the engines. The noise roused Kevin; he then pulled the throttles back hard against the aft stop. The engines spooled back to full throttle just as the right wheel went off the pavement and onto a dirt embankment. The engine thrust reversers threw up a cloud of dirt and debris. We kept going, but the momentum began to change. When we suddenly stopped, I was flung forward into the cockpit between the two pilots.

"That was close," Kevin said to Ridgerunner.

Ridgerunner, an ex-biker, was a large and intimidating man who lifted weights and walked with a John Wayne gait. He was also a very experienced Learjet mechanic. The look he gave me said that he knew what was wrong and who was to blame. "Outflow valve," he said. "The safety valve should have let go." He turned to me. "You didn't happen to mark the direction of the valve when you cleaned it, did you?"

I admitted that I hadn't. I admitted that I didn't even know I was supposed to mark the direction.

Despite his intimidation, Ridgerunner's tone never changed. Everything he said sounded as though he were ordering a hamburger, even when he said, "We're lucky the outflow valve failed on the ground. If it'd happened in flight, we'd all be dead." He took me to the back, pulled a bench seat down, and removed the interior panel on the rear firewall bulkhead. This is where I had reinstalled the safety valve less than thirty minutes earlier. "See the safety valve? You put it in backward, dipshit." He made even "dipshit" sound like "hold the onions." My ears turned red. Ridgerunner calmly explained why the safety valve couldn't prevent overpressurization when it was installed backward. He then showed me how to use a common ballpoint pen to override the pressure sensor and release the pressure. "This little pen may save your life someday, buddy." Extra pickles, please.

We were towed back to the hangar, where I reinstalled the safety valve correctly, and Ridgerunner signed off on the installation. As he was leaving, he turned and said thanks. "If you hadn't pulled the throttles back, we wouldn't have made it." What he meant was that I was forgiven for making a potentially lethal mistake—until I made another.

During this time, I was in love with Jill LaBonde, my high school sweetheart, who was now a college student in Greeley. I spent as much time as I could at the University of Northern Colorado campus. Most days, I was on the road to Denver by 5:00 AM and headed back to Greeley as soon as I got off work. I was tired all the time and, in truth, more into Jill than I was into my job. My work suffered, but I was still too cocky and overconfident to admit it.

Perhaps because of my brash confidence, I was assigned a jet for a de-mate inspection. The inspection was a big responsibility for a new mechanic, but I knew that it wouldn't be a problem—not for me. During a de-mate inspection, the wing is taken off to inspect for corrosion. The job took about three weeks, and everything was perfect until we jacked up the airplane to test the landing gear. It didn't work. We couldn't raise it.

For three days, we searched for the problem. For three days, I berated everyone who had worked on the jet. For three days, I looked for the culprit responsible for messing up my big moment of glory. Finally, the landing-gear problem was discovered—someone had clipped a tiny wire under the left wing, a real amateur move.

"Who prepped the left wing?" I shouted. Nobody answered. I asked again, louder.

"You did," Ridgerunner said.

Oh shit! I wanted to crawl under a rock. In truth, I knew the mistake had happened because I was tired and my mind was somewhere else. I also knew I had cost the company thousands of dollars and three days of hangar time. More importantly, my mistake easily could have cost lives. This experience was the beginning of my understanding that arrogance and ignorance are the twins of aviation disaster. Overconfident and complacent—I was guilty of both.

I took my damaged ego to Jill. She was a year behind me in age but light-years ahead of me in life experience and wisdom. I thought that confessing how badly I had screwed up at work would be easier if done with flowers and dinner, so I took her to a fondue restaurant called the Melting Pot in Littleton, Colorado. While I looked at an obnoxiously large menu, I said, "I had the worst day." As I told her my story, she responded with uh-huhs until I got to the part where I said, "I was thinking maybe I'm not cut out for this mechanic thing. I want to try something else."

She smiled. "Thank goodness." This was not at all what I expected. "Maybe you can find a job closer to Greeley." Jill was in her second year of college at the University of Northern Colorado, where her father, Jack LaBonde, was the wrestling coach and a professor of health education. "We could get an apartment together," Jill said.

Jill knew that I wanted to marry her. I had said so several years earlier, although it was far from a romantic proposal. It had happened when Jill's mother, a very liberal woman, offhandedly mentioned to Jill's stepfather, a former Golden Gloves boxer, that Jill had gotten birth control pills. Her stepfather grabbed me by my shirt collar. His grip was was so tight I barely had the breath to stammer, "But, sir, I want to marry her." That was my proposal. Anyway, Jill knew we couldn't live together. If we did, either her father or her stepfather would kill me.

That night, I repeated my proposal over a steamy fondue pot. Jill then calmly reminded me that her father would stop paying for her college if we got married. "If you want to marry me," she said, "you have to find a job closer to Greeley." She speared a crust of bread and dipped it in the fondue. "And you have to find a job that pays at least two thousand a month."

Two thousand dollars a month was more money than I made at Combs-Gates Learjet as an aviation mechanic. It was more money than I had ever made. "Okay," I said. "I'll do it." I had no idea how, nor did I have any idea how greatly those words and that dinner would change the direction of my life. One thing I do know as I look back—I did not make a mistake, not that night, not about Jill.

CHAPTER FOUR

Every Picture Tells a Story

"Got a helicopter out of service in Nevada, an Aerospatiale Lama," Don Guinan told me. A large man with a turned-up nose and a belly that shook when he laughed, Don was in charge of the maintenance shop. "Little John'll* take you to Mesquite to fix it." Since I was new at Airwest Helicopters, the job assignment surprised me.

Airwest offered me a position that paid enough for me to marry Jill, so I took it, despite doubts about my future as a mechanic. Our wedding was set for early fall. My new job also meant that I was now maintaining helicopters instead of jets. For me, this was like learning a new language while trying to hide a stutter. "But, Don," I said, "I don't even know what an Aerospatiale Lama—"

Before I could finish telling him that I didn't know any more about an Aerospatiale Lama than I did about a Peruvian llama, Don handed me a book and said, "Here's the manual. There's a picture on the first page." His tone left no doubt that it was useless to argue. "Don't forget your tools," Don added.

Little John, my pilot, was one year younger than I was. He stood six foot four and weighed all of 145 pounds soaking wet. He helped me carry my heavy toolbox to the airplane and set it in the backseat. I threaded the seat belts through its small handles on each side and then through my overnight bag handle. Little John had to fold up his body like a Swiss Army knife to get into the Cessna 185.

The day was calm when we left Fort Collins, but we were fighting the wind by the time we passed Denver, so we stopped for food, fuel, and a weather check in Salida, Colorado. The elevation there is about seventy-one hundred feet. As we taxied into position for takeoff, the wind, which was gusting at sixty knots out of the west, rocked the wings and lifted the tail. Most of the twelve hundred flight hours Little John had accumulated were in small airplanes in the mountains, so I was confident in his abilities. He threw the throttle forward, and we rolled down the runway. At about fifty knots, with a gust of wind, we jumped into the air. We climbed at only five hundred feet per minute. It was the best the tiny Cessna 185 could do in the winds. Below us stretched the canyon between Mount Ouray and Chipeta Mountain. To our left was the Devil's Armchair Bowl.

As we climbed, the valley climbed with us; but for every foot we went up, the valley was going up two. I looked out the window at the ground and the small streams winding through thick evergreen trees. "This isn't going to work," I said to Little John. "Get us out of here."

"Hold on tight, Dave," Little John replied through clenched teeth. "I'm going to try a chandelle." (French aviators during World War I described this maneuver as *monter en chandelle*, meaning "to climb around a candle.")

Just as I said, "Whatever it takes," he pulled the control wheel to his chest. The nose pointed straight up. For a moment, we seemed to freeze in the air, and then the nose fell off to the left, until it was pointing at the evergreen trees directly below us. My heavy toolbox behind us tugged against its restraints. I could hear the seat belts creaking. I glanced back just as the small metal handle on one side let go, and the toolbox smashed into the headliner above the rear seat.

Before I could react to the problem, I heard a loud *thud* from outside, and the airplane leaped twenty feet upward. "What was that?" I yelled.

"Trees." *Thud thud thud!* We were skipping along on treetops. As I looked out, a huge evergreen smacked against the belly of the airplane and launched us skyward as though we were on a trampoline. *Thud thud!* "If we can get one more of those, we'll be golden," Little John said. He let out a crazy laugh. Sure enough, one more bounce off a giant evergreen and we were climbing again, gaining altitude. "We better go back to Salida." He turned the wheel to the left. *Clunk!* He turned it to the right. *Clunk!* "Uh-oh," he said, "I can't turn the wheel."

Uh-oh, my ass, I thought. I heard another clunking sound behind my head. I looked back and saw that the headliner was torn and a piece of metal was sticking out of it. My toolbox had struck the roof of the airplane precisely where the control-wheel cables were routed to the bell crank and then to the wings. There was no way Little John could steer, and if he couldn't steer, there was no way he could land, and if he couldn't land . . .

"You gotta help me, Dave." Little John's near-hysterical laughter of moments ago had been replaced with a calm conviction. "When I tell you to, open your door."

"*What?* You want me to jump out?"

My reaction made Little John laugh again. "We're gonna steer the airplane with the doors."

"You can do that?"

"Maybe," he said. "Let's practice it."

I pulled the door latch, and it popped open a little. Pushing it more than a foot was like trying to carry a sheet of plywood through a hurricane. But the plane was turning. It was working. "Man, this is amazing," I said. Any fear of crashing was forgotten in the excitement.

Little John and I opened and closed the cockpit doors; Little John used the rudder and changed the power to control the pitch, yaw, and altitude of the plane. By doing this, he brought us all the way around and landed the Cessna 185 at the Salida Airport. I got out and kissed the ground.

If we had crashed, Wayne McCannon from the FAA would have reported the cause of the accident as pilot error in poor weather conditions. But that wouldn't be truly accurate. The experience with Little John had opened my eyes to something—no matter how closely you follow the rules, and no matter how carefully you prepare, things happen in flight that can lead to disaster. I also started wondering whether or not, after factoring in the unknown, it was even possible to predict the likelihood of an aviation disaster. By the time I got back home a month later, I had the beginning of an answer.

We spent the night in Salida. The next day, I repaired the bell crank on the Cessna, and we flew to Mesquite, Nevada. Today, Mesquite is a bustling oasis, a green strip of condos, casinos, and golf fairways. In 1980, it was desolate.

Terry York, a mechanic's assistant, met us at the airstrip. Little John and I were too exhausted even to help him load our gear onto the bed of the truck, a huge Ford with a five-hundred-gallon jet-fuel tank on the back. Only Terry's chatter kept me awake. "They'll sure be glad to see you," he said. "That Lama's been down for days." I reached inside my jacket and pulled out the picture to refresh my memory of what a Lama looked like.

When we reached the hotel, Terry led us around the side of a building, where each time a thirty-foot neon cowboy tilted his hat, I could see the helicopter. Sure enough, the helicopter looked like the picture. I also noticed a makeshift helipad marked out with white chalk on the new blacktop of the hotel parking lot. The whole scene was surreal. Little did I know it was just the beginning.

As I headed to the check-in desk, I heard a voice call out, "Yo! Is that my wrench?" The questioner was standing in the doorway of a trailer. "Typical fucking mechanic," the man said. "A day late and a dollar short." He tipped his cowboy hat as a hello. "I'm Dave Hodges, the only good pilot 'round here." I told him my name. "You're who? Guido Sarducci?"

"It's Soucie," I said.

He barked back at me. "Like I said, Sarducci."

In the uncomfortable silence that followed, I studied his face and posture. I was told the pilots worked long days in the hot sun, and they worked as many as forty days straight, fueled by the promise of overtime pay. He looked both exhausted and wired. I noticed the bottle of scotch he had been waving had not been opened. That's when I realized he was just messing with me. His angry face slowly transformed into a broad smile. Since I did a pretty dead-on imitation of the *Saturday Night*

Live character, I said, "Yessa, my name isa Guido, Father Guido Sarducci. You a wanna confessa your sins?"

He almost dropped his scotch from laughing so hard. Dave Hodges and I became friends that night.

The next morning, I received a first-class education in oil exploration from Tex. If I ever knew his given name, I've long since forgotten it. "I'm the HMFIC," Tex said by way of introduction. "Head motherfucker in charge. So listen up." His crumpled cap, huge rodeo-prize belt buckle, and twisted posture advertised his history as a national bull-riding champion. I listened to what he said.

The way Tex explained it, oil exploration using seismography was not a complicated process. Bundles of dynamite were delivered by a helicopter to a specified area where sensor cables had been carefully placed. The subsequent explosion allowed engineers to measure seismic activity, and these readings told them whether oil was likely to be found underground. In addition to helicopter maintenance, part of my job would be to watch the net full of dynamite that was attached to the chopper. If there were any kind of problem—for problems meant danger—I was to radio the pilot, Dave Hodges, who would "punch it off," meaning he would drop the load in a safe place. "Makes a hell of a mess if it gets dropped." Tex cocked his head and spat tobacco juice on the dusty desert floor, then continued explaining how things worked at the exploration site. "Blows the shit outta things."

My overnight trip to Mesquite somehow lasted for a month, more than long enough for me to experience the grueling life of both pilots and mechanics. We worked seven- to eight-hour days for twenty-one days straight, followed by seven days off, unless we wanted to earn extra money. Anyone who wanted extra money simply gave up his days off. That is how Dave Hodges managed to fly every day for the thirty days I was there.

In the beginning, Dave Hodges was the best helicopter pilot I had seen. Many of the pilots in the early 1980s, almost all Vietnam veterans, were unbelievably skilled, but Dave flew the Aerospatiale Lama as if it were an extension of his body. One day, I watched him chase a rabbit across the field. He mimicked every move the rabbit made, until he finally pinned it to the ground by resting the landing gear of a three-ton helicopter on its back. "Rabbit stew, anyone?" he said over the radio. He laughed and let the rabbit go. He was that good.

As Dave worked long hours day after day, I watched his ability to make good decisions deteriorate. I remember hearing from the chief pilot during one of his field visits telling the pilots, "If you get too tired, just let me know and we will get you a day off." The more fatigue I saw in Dave's eyes at the end of the day, the more I wondered how he would know when to ask for a day off. I still wonder today—why didn't somebody tell him he needed a break? Why didn't I make him stop flying?

The possibility of an accident was always present: oil exploration is dangerous, dynamite is dangerous, flying without rest is dangerous. Despite this, complacency replaced safety concerns as the crew became increasingly overconfident in their ability to perform routine events without incident. Their overconfidence began masking threats and risks, resulting in limited awareness of a potential disaster. I knew the probability of an accident was increasing daily. I worked my tired brain to identify all the things that might happen. I started writing them down: fuel, weather, lightning, human error, fatigue, and many others. Then I tried to figure out which one of the hazards was most likely to occur. I used a numbering system: fuel, 2; weather, 4; lightning, 1; human error, 5; and so on. I knew that human error was the most likely cause of disaster, but at this point, I did not know how or when an accident would happen. I felt like a psychic who knows something bad will occur but can't pinpoint a time or place. I taped the list of probabilities to the top of my battered red toolbox where I would see it every day.

During my time in Nevada, I did experience the dynamite being punched off, and Tex was right—it made a hell of a mess. I saw tall trees wrapped in dynamite and shot through the air like toothpicks. I also watched Tex go on the wildest ride of his life.

Tex always hooked the net filled with a thousand pounds of dynamite to the helicopter cable himself. When the net tightened as the chopper lifted off, Tex would slide off the net. It was all routine until the day his belt buckle caught on the net, and Dave Hodges flew off with Tex hanging on to the net of dynamite. Although I had the radio transmitter, I didn't immediately contact Dave Hodges. By now, he had gone from punch-drunk to careless to dangerous. I was afraid to tell him that Tex was part of his cargo. I was afraid Dave Hodges would drop the load, so I remained silent and watched Tex ride the dynamite net, all the time wondering if I was saving his life or ending it. Tex flailed and struggled but hung on until he reached the drop zone, where the workers untangled him. Later, I asked Dave what he would have done if I had called him. "Punch him off," he said. "Best thing to do."

When I left Nevada a few weeks before my September wedding, Dave Hodges was still flying. The terrible premonition I had while I was there came true in October 1980 on Dave's ninety-first consecutive day in the air. He wasn't supposed to fly that ninety-first day. His shift had ended on day ninety, but his replacement was late in arriving.

Late in the afternoon on that day, Dave was returning to base camp from the hills when he encountered the fickle wind that channeled through two mountains. He always called the unpredictable desert winds "the hand of God." The winds were a familiar daily occurrence, but on this day, I was told Dave cut the corner around the mountain a little closer than usual. Perhaps he was excited to get home to his wife and child and lavish them with gifts from the extra money he had earned working overtime. Witnesses to the accident said that it looked as though an invisible force simply pushed the helicopter into the mountain. At first, everyone thought Dave had been killed. Despite suffering a massive head trauma, he survived.

After our wedding and honeymoon, Jill and I visited Dave at his home in Fort Collins, Colorado. At one point during our visit, I lapsed into my *Saturday Night Live* imitation. Dave didn't laugh. He didn't react at all. He had no memory of Father Sarducci or Nevada or me. He had no memory of laughter. His body had outlived his mind.

Dave Hodges' accident showed me that complacency, a form of internal arrogance, is a root cause of many aviation accidents. Even when the most safety conscious of us do routine things over and over, we begin to stop processing the information we need to avoid accidents—information about hazard and risk, even information about the body and mind's ability to react to the expected and unexpected. At other times, the root cause is found in the failure to adequately collect and share information, what I earlier identified as ignorance. The information to avoid disaster is available, but it isn't given to the right people at the right time or in the right way. It should have been easy to predict that Dave Hodges was a risk. It should have been possible to stop him. But nobody did. The system failed. How? Why? I wanted to know.

A desire to understand the system itself and to understand the underlying assumptions and rules and regulations that define aviation became my new goal. It was as noble as it was impractical. I could not leap tall buildings in a single bound. Nobody could. At least that's what I thought until I met King Kinsey.

CHAPTER FIVE

The Odds of Surviving

Jill and I were married September 13, 1980, and enjoyed a short honeymoon in Ixtapa, Mexico. As soon as we returned home, I received my next job assignment, one that would take me out of town. Airwest had maintenance contracts for Emergency Medical Helicopter Services in Denver, Phoenix, and Las Vegas. I was sent out to relieve the resident mechanics in rotation for about two weeks in each of these cities.

Jill decided to travel with me on the four-week trip to Las Vegas and Phoenix as an extension of our honeymoon. Our first stop was Las Vegas, and Airwest provided our transportation—a rust-eaten 1970 Datsun pickup with a new paint job, courtesy of a spray can.

The small, dilapidated truck had no air-conditioning, so I filled a canvas bag with water and hung it on the rearview mirror. This was a cooling trick Uncle Louie had taught me when I was a child during the hot summers I spent in his ranch. It helped cool us just enough that Jill was able to lie on the small bench seat and rest before taking her shift on the wheel. She had finally dozed off when I saw the "Welcome to Nevada" sign at the bottom of a small hill. I pushed in the clutch and shifted out of fourth gear into third—or tried to. The entire gearshift came off in my hand. As it did, my elbow smacked my new wife squarely in the eye. We finally came to a stop, and I looked at the floor where the gearshift had been. I could see the gravel on the road through the rusty floorboards.

After two hours in the baking sun, and two hours beneath Jill's glaring eyes, the right one swelling quite badly, I repaired the gearshift, and we hobbled into Las Vegas. It was a great start to our extended honeymoon.

I had no training on how to maintain the Aerospatiale Alouette III helicopter; I didn't even have the right tools. Although similar to a Lama, it was larger and egg-shaped, with Plexiglas front windows supported by a thin metal frame. I saw the helicopter for the first time when it landed at Valley Hospital in Las Vegas. The pilot skillfully maneuvered the chopper onto the rooftop helipad with the same finesse Dave Hodges had once displayed.

Just as the wheels touched down, a door behind me sprang open. Nurses and doctors seemed to appear out of nowhere to park a gurney against the side of the helicopter with the precision and speed of an Indy 500 pit crew. Inside the helicopter, a flight nurse, with her back to me, was leaning over a patient on the lower of two stretchers stacked like bunk beds. The medical crew quickly and precisely whisked the patient away.

During all the activity, the pilot sat calmly, writing on a clipboard resting on his knee. Once he finished, he shut down the engine and pulled the rotor brake. The three large rotor blades stopped, silent. The pilot stepped out and strode toward me as though he were some victorious warrior. He was a large man with a high forehead and dark sunglasses. "I'm Paul Kinsey," he said. "They call me King." Instead of it sounding arrogant, "King" seemed fitting for the man. I introduced myself, and my hand disappeared in his as we shook. "There's a bucket and rags in a closet over there. Let's get this machine cleaned up." Somehow I knew that the "us" meant me, so I wasn't surprised when King Kinsey turned and walked to his sleeping quarters, a small apartment perched on the roof next to the helipad.

A red stripe from just below the door to the tail decorated the side of the bright orange helicopter. As I splashed a bucket of water on the side, the red stripe began to run. It was not paint; it was blood.

"Pretty gross, huh?" I jumped. I hadn't heard the flight nurse return. "We picked him up on I-15," she said. "He had a nice Harley." She told me the patient had been in a motorcycle accident, sliding over two hundred feet on the pavement. "There wasn't much left of him when we got there." She opened the door, and more blood ran down the side of the helicopter. "Gotta get ready to go back out," she said. "We need to start in here." So I did.

A week later, with the blood and gore of emergency medical transport now familiar, I took the elevator to the rooftop after stopping off for coffee and a dose of hospital food. The helicopter was on a trip, so I began preparing the tools I would need for the daily greasing when it returned. I heard sirens and saw an ambulance roll out of the garage below and into the street. When I looked up again, a dark wall of smoke was building over the Las Vegas skyline. Then I heard the helicopter on its approach and jogged out of the way. The sound of the blades told me that the helicopter was approaching faster than normal. Even so, King Kinsey set it down gently on the rooftop. Unlike the first time I had watched it land, I could not see through the Plexiglas bubble this time. The window was as black as that of a tinted limousine hiding its occupants.

As soon as the helicopter landed, the cargo door opened. A man and a woman who had their backs against the door fell out onto the helipad. Several more people frantically jumped over them to get out of the helicopter. The number of people on board clearly was more than the chopper was licensed to carry. Once they were all on the rooftop, no one seemed to know where to go or what to do. Nurses appeared and tried to guide them inside, but several people wandered around, dazed, walking dangerously close to the whirling tail rotors. That's when King Kinsey slammed on the rotor brake and seconds later ran toward me. "Dave, we need to get this window cleaned up so I can go back. There's a lot more of them up on the hotel roof."

The day was Friday, November 21, 1980. Little by little, I heard the story. The MGM Grand—a twenty-six-story, two-thousand-room casino hotel—was on fire. Some five thousand people were there when the fire started. Some were trapped inside, while a growing crowd had made it to the rooftop, waiting to be rescued. That's what Paul Kinsey was doing. That's why his Plexiglas window was black. That's why the helicopter was overloaded. That's why I needed to figure out a way to protect the window so he could see to fly.

Besides Paul Kinsey, two other helicopter pilots were ferrying people to safety—Ray Poss, who was co-owner of Silver State Helicopters, a charter business, and Las Vegas police pilot Harry Christopher. The scene that awaited them on the rooftop of the MGM Grand was chaotic, as people abandoned civility in favor of survival instincts. The chaos meant increased danger for the pilots as, one at a time, they landed on the MGM rooftop, filled their helicopters with people, then flew through blinding smoke, descended, and unloaded.

After a few trips, the flames had melted the window of the big Aerospatiale Alouette III so badly that Paul Kinsey could no longer see. Each time he unloaded, I sanded the black soot off the window and rubbed it with Vaseline to protect it from the heat. My efforts worked long enough for Paul and the two other pilots to extract hundreds of people from the towering inferno. The hardest part of the day, one of the pilots later said, was when he was hovering below the rooftop, where he could see all the people trapped inside, leaning out of the windows, begging to be saved, and knowing there was nothing he could do.

The next day, the local television news reported eighty-five deaths, including fourteen firemen. Eighty of the deaths were from toxic-smoke inhalation. More than 650 were hospitalized.

Smoke dampers were built into the ventilation system and were designed to prevent the circulation of smoke in the event of a fire. Tragically, they failed to function properly and allowed toxic fumes and smoke to circulate throughout the hotel.

The report of failed smoke dampers echoed in my head for several days following the tragedy. The designers of the building designed a mechanism to prevent death from toxic smoke. They recognized the threat of fire and the risk to life associated with smoke and fumes. The risk was high enough to build the smoke dampers. So why did they fail? Was it a poor design? Was it a lack of specific routine maintenance? Or was it a failure to maintain the dampers in the way the engineers recommended?

I was appalled that the threat of fire and toxic smoke was recognized and addressed by the designers, but ensuring the smoke dampers were functioning properly wasn't important to the hotel. I speculated that the hotel management must have seriously underestimated the likelihood of a fire occurring or the number of tragic deaths from smoke inhalation. Why else would they have allowed the dampers to be nonfunctional? Was it too expensive to maintain and test the dampers, or was fire simply not considered a credible threat?

I then began to wonder how the hotel management would have prioritized the maintenance and testing of the smoke dampers. There are plenty of formulas for return on investment, but was there a formula for calculating risk against loss of life? How could the risk of operating a hotel without maintaining smoke dampers and the probability of loss of life be weighed against dollars and cents?

I had no answers, neither on that day nor a few years later, when I became director of maintenance for Air Methods and faced the same question when Mike Myers asked me to install wire-strike kits. How could the potential risk of flying without wire-strike protection and the probability of loss of life be weighed in dollars and cents? Even so, on the day of the MGM Grand fire, I had the first inkling that the disaster was caused by decisions made long before the event. I also wondered, perhaps for the first time, if the underlying cause of aviation accidents followed a similar pattern. These thoughts about evaluating risks and hazards and safety would preoccupy me for the remainder of our extended honeymoon trip as we went on to Phoenix.

After Phoenix, we returned home to Colorado without further breakdowns or tragedies. We were now a few months into our marriage. With marriage came changes, some expected and some unexpected. For one thing, Jill changed her major from theater arts to business. She really enjoyed the people in the business classes, and a business major offered more opportunity and earning potential. I sometimes studied with her at our apartment. The case studies and the strategic analysis tools she used for investment and financial risk management fascinated me. There was a direct parallel between the case studies of businesses that failed to manage risk and my observations of air operators who had failed to manage risk. I could also see how the strategic analysis tools could be adapted and used in aviation to measure the likelihood of an occurrence, and weigh it against the impact to prioritize multiple risks. I knew this realization was an important piece of the safety puzzle, but I also knew that until I understood the underlying cause of aviation accidents, I would not save a single life.

For the next year, I traveled to Phoenix and Las Vegas several times as a relief mechanic while Jill finished college. I never once returned to Las Vegas without recalling the fiery November day, with me on the rooftop frantically sanding the smoke-black windshield to regain visibility so Paul Kinsey could ferry people to safety. I also never forgot how decisions made long ago and usually from far away sealed the fate of so many people.

As it happened, I was not in Las Vegas, or even still working for Airwest, when I heard about Paul Kinsey's death. His helicopter crashed on December 7, 1983, in Black Mountain, Nevada. He had been on an air-ambulance flight along with two flight nurses. All three people on board were killed. The National Transportation Safety Board (NTSB) investigation blamed Paul for the crash. In the air or on the ground, when dealing with aviation, it is easy to go from hero to fool. Whatever really happened on Black Mountain, Nevada, Paul's crash proved to be an eerie foreshadowing of Mike Myers's fatal accident on Christmas Eve two years later.

Of course, at the time, I knew nothing of the tragedies lurking two years in the future. All I knew was that my own future was about to change dramatically, for in 1981, I decided to trade my toolbox for law books.

I decided to become a lawyer.

CHAPTER SIX

The Sisters of Charity

I enrolled in pre-law classes the same year. I was convinced that studying law would provide answers to my questions about how to improve aviation safety, and even an answer to the question of how much control I had over my own life. My employer, Airwest, was supportive of my plans and offered me a job that was neither demanding nor involving travel. Airwest wanted to provide emergency medical helicopter service for Saint Francis Hospital in Colorado Springs. My new job would be to maintain the helicopter.

Bob Theidean, both a fine pilot and the sales and marketing manager, flew an Alouette III helicopter from Fort Collins, Airwest's headquarters, to Colorado Springs the night before I first saw it. "Our contract was just awarded yesterday," he told me, "so the pilots won't be here for a few days yet." Bob's mustache was turned slightly upward and animated his usual welcoming smile.

"How did we get a contract so fast?" Airwest seemed even less prepared than usual, which was saying a lot.

"We didn't," he said. "Sister Mary* awarded the contract to Air Methods, Roy Morgan's new company." Bob's smile reappeared. "But there's a clause in our contract that gives us dibs. Any hospital run by the Sisters of Charity has to offer us the EMS first." Bob raised his arms in victory. "Home team scores again."

Something about the story bothered me, but I didn't let on. I simply told Bob that I had to leave for class. My first aviation law class started soon. Bob pulled out some papers. "Drop this contract off with Sister Mary, will you? She needs to sign it." I took the papers and hurried off.

I used a shortcut through the hospital to reach the administration offices. When I neared Sister Mary's door, I heard someone say something about "the helicopter." I stopped to listen. "It's preposterous," Sister Mary said. She was a slight woman, but her voice carried authority. "They can't do this to me," she said. "To you!"

A male voice said, "It's all right, Sister." My eavesdropping turned into blatant espionage as I stepped behind the door to listen. It was unusual to hear agitated voices at the hospital, or even to hear a man's voice. My curiosity made me peek inside. A man stood opposite Sister Mary. He held both of her hands in consolation. "They're doing what they think is right." A moment later, after saying good-bye, he picked up his coat and walked toward me.

"I won't forget this, Roy," Sister Mary said.

Until that moment, I didn't realize that the man I was spying on was Roy Morgan. I stepped out from behind the door. As he passed me, his eyes flickered to the Airwest patch on my shirt and the contract in my hand. They clearly marked me as the competitor who had robbed him of his opportunity. He smiled and said, "Good day, sir."

"Can I help you, son?" Sister Mary's voice was flat and businesslike when she spoke to me. It reminded me of the knuckle-rapping nun at Saint Joan of Arc Sunday School.

"Bob asked me to have you sign these," I muttered.

She snatched the papers from my hands and flipped through them page by page. "If you want to amount to anything in life, son, be like the man who just left here," she said. That's when I noticed another contract lying on her desk. It was ripped in half. "When I told Roy we couldn't honor our contract, you know what he did? He didn't get mad or threaten to sue us. Instead, he just said for us to call him if we needed his help in the future." Sister Mary dropped the torn-up contract in the trash and invited me to sit.

As soon as she had settled in her chair and tucked her hands in the front of her robe, she started telling me about Roy Morgan. If her story was long, I would be late for my pre-law class. I fidgeted and looked out a window at the Alouette III. "Roy was a helicopter pilot for Public Service," Sister Mary explained. "Thirty years he worked there. Anyway, one day Roy and a friend went out hunting, and the friend got hurt. Bad. Shot accidentally. So Roy covered him up from the cold and climbed to a ridge to contact mountain rescue with a handheld radio. Then he went back to his friend. But the radio reception was bad, and he had trouble guiding the helicopter to their location. Because it took so long, his friend died during the flight out. Roy was devastated."

She stopped and withdrew her hands from her robe to wipe off a tear. A moment later, she cleared her throat and continued. She told me that after the death of his hunting partner, Roy Morgan mortgaged his house, found a friend who did the same, and started Air Methods. Even though he was nearing retirement age, he also went back to college to study business management. He was that passionate about improving emergency medical service. While Sister Mary was talking, I noticed it was 2:15 PM. I should have been sitting in class. Maybe I was, in a way.

"Did you know that over half of all critically injured patients brought in by helicopters die during the trip," she said, "just like his friend? And another 25 percent die within a week of making it to a hospital." I was surprised to hear a nun talk of death in statistical terms. I flashed back to my chart of hazards and risks in Nevada. "So Roy said"—she brought her chin to her chest and imitated a manly voice—"we need to bring the hospital to the patient and not the patient to the hospital." She had my full attention now. "You see, Roy's helicopters have special medical kits to provide treatment on site. He says it'll reduce transport fatalities by 35 percent and post-arrival fatalities by 40 percent. And I believe him."

I suddenly realized that she wasn't crying because Roy Morgan unfairly lost a contract. She was crying because she believed that without Air Methods' emergency medical service, people were going to die needlessly.

I didn't like what had happened, but I still had a job to do. If anything, my discomfort made me try to do it even better. Sister

Mary's story also fueled my growing interest in aviation safety. I shared my concern about business decisions negatively impacting aviation safety with my professors at Colorado College, but my concern fell on deaf ears. One professor, after listening to the story about Airwest outflanking Air Methods for a contract, even though Air Methods was better equipped to save lives, responded by saying, "Well, the law isn't supposed to be fair." *If that's true*, I thought, *why am I studying law?*

Jill picked me up from school that night. When we entered our apartment building, we heard our phone ringing. It was after 9:00 PM, so I expected the caller to be someone from the hospital reporting a grounded helicopter. I ran up the stairs to answer it. "Dave's hangar. You buy 'em and fly 'em; we get 'em and fix 'em." All the dispatchers had a good sense of humor. I waited for a laugh; there wasn't one.

"May I speak with Mr. Soucie, please?"

I was still panting from the run up the stairs, but I managed to spit out, "This is he." Jill brought me a glass of water, so I covered the receiver and took a drink.

"This is Roy Morgan with Air Methods." I choked on the water and then spewed on a new fern in a macramé hanger.

I thought, *Why is he calling me?* Did Sister Mary give him my phone number? *Is he going to rip into me for Airwest's stealing his contract? Why me? I'm only twenty-two years old. I'm not the boss. It wasn't my idea.*

It didn't take long for Roy to answer my unasked questions. "I'm looking for a director of maintenance," he said. "I'd like to talk to you about the job." Holy crap! Roy Morgan wanted me! I stammered something, and Roy invited me and Jill to his house for dinner the following night.

We arrived at a modest house in Aurora, Colorado. Once introductions had been made, Roy offered us drinks. Jill and I had little experience with cocktails and social niceties, so we both said, "Just a beer."

I had tripped coming up the front porch, and when I sat down, I tried to cover the knees of my trousers. I had fallen just as Roy answered the door. Fortunately, Jill looked great in a blue velvet skirt and high heels.

Roy got down to business fairly quickly, telling me about the job at Air Methods. He said the salary was twenty-seven thousand dollars. He even apologized for not being able to pay more. "We're a new company, just starting out," he said. Jill and I looked at each other and stifled our glee. The pay was three thousand dollars more than I was making at Airwest. "By the way, do you smoke?" Roy suddenly asked. I saw his wife at the bar opening a small oriental box like the one some of my friends used to stash marijuana.

"Not really," I told him. "I haven't smoked pot since high school."

Roy and his wife looked at each other as she pulled the cocktail napkins out of the box. "I meant cigarettes," Roy said softly. He was anti-smoking.

Roy called me the following day. In spite of the previous night's gaffe, he offered me the job as director of maintenance. The job with Air Methods meant giving up school and my plan to study law. A few months earlier, I would have said that nothing could entice me to do that. But a few months earlier, I had not met Roy Morgan. I wanted to save lives; I couldn't say no.

A week later, I entered Air Methods' offices located in a suite at Arapahoe airport in Aurora. The walls were bare except for a large map of the United States. Colorado had three tiny colored pins stuck in it: Grand Junction, Greeley, and Aurora. Instead of a pin, there was a small hole in Colorado Springs, the only remaining evidence of Roy's ill-fated contract with Saint Francis Hospital. I felt good about my new job.

The front offices had mahogany desks and bookshelves. They were simple but elegant. *Nice*, I thought as Roy escorted me to the first cubicle. "Here it is!" he said. The director of maintenance office was nothing more than an empty cubicle. There was no desk, no chair, and no bookshelf. The only wood was the handle of a squeegee poking out of a mop bucket. "Any questions, come and get me," Roy added. "I'm just down the hall." He turned and walked off.

Jill and I bought a small house not too far from the office in Aurora. Our son, Tyler, was born in 1984. I was twenty-five and had a beautiful wife and a healthy son. I knew my fortunes would rise along with the company, so I worked long hours, often neglecting my wife and my son.

Air Methods' success came so fast that Roy could hardly keep up with the demands. When you can't keep up, you cut corners, the way Roy did with the "med kit" that appealed to Sister Mary. The med kit might "bring the hospital to the patient," but it was illegal. It had not been certified. I redesigned the kit to meet FAA standards and got it approved. This simply increased my workload.

Roy had a great sales pitch for helicopter services. Hospitals in remote or thinly populated areas that relied on road ambulances could draw critically ill patients from about a twenty-five-mile radius. Because of a limited number of patients, these hospitals found it very expensive to afford a specialist like a heart surgeon. But if the service radius were enlarged, there were more potential patients and more money to pay for expensive specialists. Roy showed hospitals that they could expand their service radius to 250 miles by using helicopters to transport critical patients. This was how he sold hospitals on signing contracts with Air Methods. Everybody won.

The company grew quickly. By the time I left, five years later, Air Methods was the world's largest emergency-medical-services helicopter company, with fifteen helicopters and two fixed-wing aircraft, a Learjet and a Beechcraft Baron.

Our contracts were not limited to Colorado either, and we frequently traveled to Oregon and Texas. I often accompanied Roy in the Beechcraft Baron. Other times, I flew with Rob Hass*, our avionics expert. I still had no pilot's license, and although I became fairly proficient, I flew in the second seat.

In late December 1985, Rob and I were flying home to Aurora after a visit to Bell Helicopter in Dallas. During the flight, clouds had built up so thick that we were bouncing around like five-year-olds on a trampoline. After a while, Rob said, "Didn't you check the weather?"

"Was I supposed to?" I said, although I knew he had asked me to check the weather conditions. Tired and thinking about my neglected family, I had totally forgotten to check the weather. I just wanted to get home for Christmas.

We flipped left and right while we searched for a safe route through the clouds. A ray of sun caught my eye, and I directed Rob toward it. After a stomach-jumping descent through the black wall, I looked out the window for a familiar sight. I finally saw a water tower as we dropped out of the clouds. The name *Salina* was painted on it.

"Kansas?" I said.

We landed in Salina, Kansas, just before the blizzard struck. For two days, we watched it snow, living on candy bars, potato chips, and Cokes from airport vending machines. Two feet of snow covered the runway. My mood was as grim as the weather. Missing Christmas was not going to do anything to smooth over my neglect of Jill and our son. I knew I had to do something to change the hours I was working and the stress I was experiencing. Then the weather broke, and I thought the gods had smiled on us. Little did I know that behind the smile awaited a grim fate. I made it home just in time for Christmas Eve. The year was 1985.

It was the same Christmas Eve that a blizzard swept across the western states, the same Christmas Eve that an Air Methods' helicopter was sent out to a Utah clinic with two nurses to help deliver a baby, the same Christmas Eve that the helicopter crashed after hitting a power wire.

It was the Christmas Eve that Mike Myers died.

CHAPTER SEVEN

Flying Blind

As Jill and I passed through the reception area for Mike's memorial service, people were warm and kind to us. It felt strange. At the least, I was expecting some awkward glances and feet shuffling. Surely, I wasn't the only one to make a connection between my decision not to install wire-strike kits on our helicopters and Mike's death.

We were late, and the service was already under way. We stood beside one of the doors in the rear of the gymnasium. Roy Morgan was at the podium. He was talking about Mike's wife and beautiful children, his heroism, his devotion to God, and his dying doing what he loved. When he finished, Roy invited anyone who wanted to say something about Mike to come forward. I must have shifted or stepped forward because Jill touched my arm. "You sure about this?" she whispered. Jill had always been good about saving me from myself. She knew it was the wrong place and time for me to tell the world I was sorry.

I had barely settled back against the wall when I heard, "You can't blame yourself." Steve Bloomquist, the chief pilot for Air Methods, stood beside me. He whispered, "You did what you were supposed to do, what any good executive would do, what was expected of you."

Steve's words were meant to relieve my distress. Instead, they made me realize that nobody there blamed me for Mike's death, not because they were unaware of my role in it, but because they simply accepted that I was doing my job. It was business as usual. *God help us*, I thought.

The next day, I settled behind my mahogany desk and leaned back in my leather chair and studied the walls filled with certificates for the training I had received. None of my certificates was for training on how to improve the aviation world. Why was this so important to me? Did I merely have a Superman complex? Was I alone in my belief that aviation safety and business profit conflicted? These questions remained unanswered when I attended the FAA investigation into Mike's crash a few days later.

Federal regulations require that every accident resulting in death or serious injury be investigated. While the NTSB is responsible for the investigation, unless the accident is deemed significant, that is, widely covered by the news media, it often delegates the responsibility to the FAA. The investigation of Mike's accident was led by FAA Inspector Gary Gomes. I met him at Air Methods' offices in Grand Junction, Colorado, in January 1986. Gomes sported a sharp suit and an even sharper persona. He got right to the point. "Far as I can see, it's pilot error. In that kind of weather, he should never have taken off."

I couldn't hold my comments back. "But if Mike didn't at least try, then the patients—the pregnant woman and her baby—would have died, right?"

"Maybe so, but that's not what we're here to investigate," he said. "We're here just to figure out the probable cause of the accident. That's what I'm doing."

I shook my head no. "If he'd had a wire-strike kit on the helicopter, he wouldn't have crashed." I recently had seen a demonstration of a wire-strike kit. I was now an avid proponent of it. "It would have saved him," I said.

Gomes put his hand on my shoulder. "He still could have hit the wire from the side. You don't know that anything you did could have saved him."

I was stunned. "So there's no violation? Nothing?"

"No violation. No follow-up. It's not necessary," he said. He got up to leave. Meeting over.

"What are we supposed to do?" I said.

My question got his attention. "What do you mean?"

"Well, what do we do now? How do we know that other helicopters aren't going to hit wires? Even if I put wire-strike kits on our helicopters, what about the other companies? All the other people flying at night?"

"They'll just have to learn. Same way you did."

For a while, I put the issue aside. I was a convert, but I didn't know how to convince Air Methods to spend the money to install equipment that wasn't required. I longed for a change in FAA regulations that would force us to add wire-strike kits, but after talking to Gary Gomes, I knew that wasn't going to happen. I had to find another way. The last place I expected to find a solution was at the Helicopter Association International (HAI) convention. I attended the conventions frequently, but this time, I paid particular attention to the display booth selling wire-strike kits. This time, I went around talking to people, saying, "Did you see the wire-strike-kit booth over there?"

Everybody would say, "Yeah, yeah, yeah."

"You know, we had a fatal accident because we didn't have the kit on our helicopter."

"Yeah," the other person would say. "Us too."

That was the whole discussion. Nobody ever asked me if Air Methods was going to install the kits. Nobody ever said, "We're doing it" or "We need to put safety first."

Somebody had to step up, so after the convention, I asked Roy Morgan for the money to install the kits on all our helicopters. "We don't have it," he told me.

"Other people are going to die," I said.

"I don't know why you feel responsible for what happened to Mike," he said. "The FAA cleared us." Roy's response disappointed me, even if what he said was true. The company, despite its growth, was scrambling for money, and the FAA report had cleared us and blamed Mike.

"Well, we have to find the money," I said, "because I'm not comfortable letting our pilots fly at night." I rarely stood up to Roy in this way. He knew I was dead serious. As we stood there, I could almost hear the gears in Roy's head as he struggled to crunch the numbers the way my new Apple, Lisa, did when it made annual cost forecasts. He was evaluating cost versus value: how long would it be before somebody else died in a wire-strike accident? If someone did, would the company be blamed? His expression told me that I was losing. Then I remembered my conversations at the HAI convention, and I knew how to sell the idea. "Look at it this way," I said. "If we install them before our competitors do, we can use this to sell to the hospitals. We'll be the safest helicopter service, the one better equipped to fly safely at night, so we'll get a bigger market share. It's good business." Roy's expression changed as he saw profit instead of loss. He was on board.

Later, I presented this same line of reasoning to the board of directors. Everybody liked it; everybody liked beating our competitors. I learned a valuable lesson in management—the single best way to improve aviation safety is to tie the improvement to profit.

Because I was able to tie safety to profit, Air Methods became the first emergency helicopter service to install wire-strike kits. I got what I wanted, but at a cost. I could no longer pretend that Air Methods and Roy Morgan were on some mythic journey to improve the world. Any remaining illusions I still had were squashed later that same year when the first baby went blind.

In 1986, Air Methods began offering a new service—transporting newborn babies in incubators. Since neonatal transport required a specialized team to treat a baby inside the incubator, weeks of training were needed to prepare a team for this life-saving service. After all the preparation, we were finally up and running.

One of our first calls was from the small mountain town of Frisco, Colorado. The neonatal team we transported skillfully handled the delivery, but the premature baby required specialized care at a hospital. An incubator was strapped to a stretcher and placed in the helicopter. Walt Wise was the pilot that day. After taking off, he headed east along the I-70 corridor to Denver. Thick clouds were building over the mountains. Despite the clouds, Walt could see a line of cars on the highway. Sunday-night traffic on I-70 on a ski weekend turns an hour-and-a-half trip into one of three to four hours. Ground transport was out.

Walt decided to fly higher to see if he could avoid the turbulent weather. He finally saw a path around the storm, but it was at fifteen thousand feet. A helicopter is not pressurized, so he informed the flight nurses to prepare for the high altitude. Everyone donned an oxygen mask—everyone except the baby in the incubator. Nobody yet realized that the designer of the incubator had made an operational assumption that would be disastrous.

Years before medical-helicopter transport existed, the designer of the incubator had assumed that ten thousand feet was the highest altitude at which it would be used. As such, while capable of managing the critical oxygen-nitrogen mixture from sea level to ten thousand feet, or in pressurized airplanes, the incubator was incapable of working properly at higher altitudes in nonpressurized helicopters.

The flight was harrowing, as Walt took a zigzag route to dodge thunderstorms and lightning bolts; the trip to Presbyterian/St. Luke's Medical Center in Aurora took an hour. Upon landing, the flight nurses congratulated themselves on a successful mission, while they rolled the incubator into the neonatal-unit critical-care room. The doctors examined the baby. They discovered he demonstrated signs of retinal damage and slight blindness.

At first, everyone assumed that the child had been born with limited sight, but several weeks later, a second premature baby transported to the hospital from the mountains also arrived with signs of blindness. This was far too coincidental.

The connection between pure oxygen and infantile blindness was discovered in the 1800s. Since then, a safe mixture of oxygen and nitrogen has been used to prevent the effects of pure oxygen on the retina. The risks of supplemental oxygen had been all but forgotten. Somehow, air transportation of infants had created a new threat contributing to retinopathy of prematurity (ROP).

After a frenzied investigation, we discovered that the regulator in the incubator, when used in a nonpressurized helicopter, compensated for the changes in altitude by increasing the oxygen mixture. At very high altitudes, the oxygen saturation was enough to damage the retina of an infant. The flight nurses had the ability to manually decrease the oxygen level, but they had no indicator to tell them when the oxygen had reached a critical level. The result was tragic.

I was determined to do something, but it took days of trial and error before the solution appeared in the form of a 1975 Datsun 510. I had driven the Datsun while in school at Colorado Aero Tech. Since then, it had provided white-trash ambience in a field beside my mother's house. One day, I remembered that when I drove the Datsun from the foothills to the mountains, the carburetor automatically changed the fuel-to-air mixture to compensate for altitude.

I grabbed my toolbox and headed to Mom's house. Once there, I carefully studied the carburetor before removing the compensating bellows from under the throttle assembly. "Just what I need!" I shouted. My outburst and jump for joy ended in a collision with the hood of the car. I retrofitted the bellows invention into all Air Methods medical kits. My invention was also incorporated into all our future designs. There would be no more babies blinded by helicopter flights. Not with Air Methods. Word of our success spread, and other companies soon heard of my invention. One of these companies was SFENA, a global aviation and avionics company in Sweden.

I met the vice president of SFENA at one of the frequent Helicopter Association International conventions. He was a tall Swede with a professional, but approachable, demeanor. I assumed he was a potential buyer of our medical kits when I ushered him into the shell of a Bell helicopter that Air Methods had pilfered from a boneyard and outfitted with a full medical interior. We had trucked the helicopter carcass to the Chicago convention, where it was the center of attention. We sold the medical kits to Americans, Europeans, Saudis, French Canadians, and others. Because of this, I was not surprised by the interest of a SFENA executive until I heard his questions. They weren't the usual ones. Roy quickly took the Swede off to the enclosed sales area, and I finished our presentation alone.

That night, Roy informed me that SFENA wasn't looking to buy med kits from us; instead, it wanted to purchase the rights to manufacture them. As an employee of Air Methods, I did not own the patent on anything I invented, so I had no power to decide this. Roy had the power to say yes or no to SFENA. He asked me to go to the SFENA offices in Texas the following week to negotiate the transaction.

When I arrived in Dallas, an entourage of SFENA representatives greeted me. They were eager to learn all about our medical kits. I explained and demonstrated, negotiated and signed documents, and then headed home.

When I arrived back in Denver, Roy was in a meeting with potential clients from a hospital in Bend, Oregon. He saw me standing outside his door and excused himself. He told me that we needed to start on the design of a brand-new med kit, a new kit designed for a Bölkow Bo105 helicopter for the hospital in Bend. I studied his face. Had he forgotten why I went to Dallas?

"We can't do that," I said. "I just signed a non-compete agreement with SFENA."

Roy smiled and said, "Yes, *you* did. But I didn't."

It took me a moment to understand what I was hearing. I stood dumbfounded as Roy turned his back and said, "Get started, Dave. You have a lot of work to do."

From that moment on, it no longer mattered to me whether I had been fooled by Roy from the beginning—his noble gesture with Sister Mary could have been merely a shrewd business strategy—or whether Roy truly had been a good man who had buckled under business pressures. Either way, he was asking me to trade my integrity for the profit we would make manufacturing medical kits. Either way, my belief in Air Methods as my route to improving aviation safety had been crushed. Either way, I had been duped. I began looking for another job the following day.

CHAPTER EIGHT

White-Collar Grime

The first person I contacted about a job was Hans Hilkuysen. In 1980, Hans was the man who had hired me at Airwest. Since then, he had moved on to Omniflight Helicopters in Janesville, Wisconsin, where he managed all its profit centers. Omniflight profit centers were independently operated sites for maintenance or charter service or both. Hans told me that the profit center in New York was looking for a new manager. I flew to the Big Apple on a weekend to avoid alerting Roy Morgan that I was job hunting. For a full day, Hans and I rode around with the pilots and met the site managers and maintenance staff. By the end of the weekend, I had the job. The management position meant I could put away my toolbox and try out different skills. I rushed back to Denver to tell Jill.

I walked into the house and could see that Jill was excited. I thought she already knew about the job offer, although I couldn't figure out how. Instead, she said, "I'm pregnant again!" I was thrilled; we both were. But Jill was also concerned. She had been told that carrying the baby full term might be difficult. Whatever decision we made, her pregnancy had to be carefully considered. We weighed our dislike of raising a family in New York City against our dislike of my remaining at Air Methods.

On Monday morning, I told Roy that I was leaving. "How can you do this to me?" he said. I don't know what I was expecting—anger perhaps—but Roy appeared genuinely distraught. "I really don't want you to go," he told me. "I need you. We can't build the med kits without you. Just tell me what it'll take to keep you at Air Methods."

Maybe my ego was swelling. Maybe I still liked Roy, despite everything. Maybe I was concerned about raising Tyler and a new baby in New York City. I decided to negotiate instead of walk away. "To start with, I want to be made a VP," I said. "And I want a substantial raise. With the new baby and all, I need it so we can get a bigger house. And I want a percentage of the sales of the med kits." I waited for Roy to react. He didn't. "I'd like a new van too, instead of that little thing I'm driving."

When Roy still didn't respond, I started to walk out of his office. We were New York–bound. "Done," he said. I couldn't believe it. After years of being awed by him, I finally beat Roy Morgan. Maybe that's why I stayed; I wanted to experience the feeling that I had bested Roy.

The following week, Jill lost the baby. We were disappointed but didn't yet realize what it truly meant. To stay or go now seemed minor compared to our family crisis, but when I finally called Hans, I still lied to him. I told him Jill was pregnant and that the stress of moving and changing jobs was too much for her. As a young man, lying was easy for me. Put me in a jam and out popped a lie. Oddly, it didn't occur to me that the road of dishonesty I sometimes traveled made me just like Roy.

Roy delivered on some promises and reneged on others. As usual, one thing he always rewarded me for was getting the job done. I worked hard and drove others hard. In truth, I was a workaholic. Roy loved it, but Jill was less thrilled by my obsession. With each passing year I worked at Air Methods, I ignored my wife and my child more. At one point, she even found a Workaholics Anonymous group and suggested I attend. I knew Jill was right, so I went to the next meeting. Nobody else showed up. It was like a cosmic joke. Workaholics were too busy working to attend a Workaholics Anonymous meeting.

While I was designing the new med kit for the emergency medical helicopters in Bend, Oregon, I still had my duties as director of maintenance. These months were a particularly busy and difficult time for me. We had bought two new Bell 222 helicopters. "Engine Problems Guaranteed" should have been stenciled on their side.

It was no surprise when Roy sent me to Minneapolis to oversee an engine overhaul. "We need the Bell back in the air immediately," he had told me, so for two days, I worked without leaving the hangar. The mechanics and I ate and slept whenever and wherever we could. Despite this, the helicopter wasn't ready. I pushed even harder, and the crew finally rebelled. They wanted a full night's sleep and hot food; they wanted their wives and kids. No matter, I kept pushing them until we got it overhauled and flying.

When I got back to Air Methods, I was expecting congratulations, if not a reward. Instead, I was fired. "People are afraid of you, Dave," Roy said. "People say you drive them too hard, push them too hard, and you won't listen to reason. They're afraid you're going to do something that'll cause a disaster. I can't have that."

I was stunned. The actions Roy used to justify firing me were the same actions he had rewarded for five years. My shock was so great that when he said there were other positions at Air Methods for me besides director of maintenance, I didn't hear it. I just nodded and stumbled out of his office.

It didn't take Sigmund Freud to figure out that Roy was a replacement for the father who had abandoned me. Maybe that's really why I had stayed at Air Methods. Maybe that's why being fired hurt so much, even though I had made a pact with the devil when I bartered away my integrity to stay. Only a fool thinks the devil won't collect. What had happened to me? Where was the man who had suffered to watch Dave Hodges destroy himself? What had happened to the inventor of the bellows kit to save babies' sight? Was I still the man who wanted most of all to make air travel safer? I didn't know anymore; I had to find out.

New York was no longer an option, but Hans still wanted me to work for him. He told me the Chicago profit center wasn't making enough money and was rife with problems. He offered me the job as manager. I took it.

Chicago was a disaster from the beginning. I had envisioned my family in a beautiful house near Magnificent Mile, a private school for Tyler, an advertising job for Jill, and a new company car for me. Instead, with the mortgage rate at 18 percent, we couldn't sell our Denver home and had to rent it. With little money to buy a new house, we settled for a run-down, cockroach-infested tri-level home in the woods south of Chicago. Tyler was enrolled at a small public school. Jill couldn't find a job. My company car was an old Chrysler K-car missing a muffler. On top of all this, at work, I found a second-rate team of pilots and mechanics.

As manager, I was in charge of eight mechanics, six pilots, and an office staff, plus three Bell LongRangers and ten Hughes 269 helicopters. I had inherited the equipment and contracts, as well as the personnel, from the previous manager. It was my first executive position, and I wanted to make it work. It was not going to be easy.

To start with, my secretary betrayed me. Once, after I sold some test equipment I owned to the electronics shop, she secretly informed Hans that I was bilking the company. Hans told me to fire her, but I kept putting it off, even though she was spreading distrust and discontent among the employees. I was not good at firing people. Several weeks went by, and I did nothing to stop the mutiny under way in Chicago. Not only was my secretary trying to get me fired, but the mechanics came to work drunk and showed each other the guns and knives they kept in their toolboxes. The day the head mechanic fired his pistol at a bird that flew into the hangar, I knew I had to act. "Put the gun down!" I yelled.

"Or what?" he said with a snicker.

"Or you're fired."

"Man, you don't have the balls."

Rationally, I knew that firing him while he had a gun was not all that bright. I looked around the hangar at the fuel and oil barrels, the cleaning solvents, and the helicopters. Another shot could become a death sentence.

"You're fired!" I said. I took the gun out of his hand and told everyone there to remove any and all weapons from their toolboxes or they, too, would be fired. I felt like the sheriff marching into the saloon to disarm rowdy cowpokes. Weapons started coming out of every toolbox and most coat pockets. The look on their faces told me they were torn between handing them over and using them on me. My heart was beating fast. Finally, a young mechanic started gathering up all the weapons. I breathed a sigh of relief and sent everybody home for the day.

Then I turned to my secretary. "Hans told me about you. Get out! You're fired!" She silently gathered her stuff and followed the mob out the door. I never saw her again. Things settled down after this, but the shop was never right for me. My first white-collar job was proving to be nastier than wielding a grease gun.

Despite all the problems, I stayed with the job until the owner became incapacitated with a terminal illness. Instead of the company promoting Hans to run operations, they hired a Navy admiral with no commercial aviation experience. After one meeting with him, I quit.

We decided to relocate back to Colorado. I took a drive near Jeffco Airport one day and saw Colorado Aero Tech perched on the hill. I parked my car just across from the school to take in the view of the mountains and reminisce about the time I spent there. I had no plan, so I just sat there. Across from the Colorado Aero Tech was a lighted sign for Turbowest. Turbowest was a Beechcraft and Piper Aircraft dealership with a large maintenance shop. I remembered that Doug Stimson, an acquaintance, was the general manager. I decided to stop in to see him.

"Soucie!" Doug bellowed. When Doug was around, everyone knew it. "When're you gonna come work for me here on the dark side?" Dark side? He had no idea what I had been doing. I told him I was open to an offer. "Well then, let's go talk to Finnoff right now," Doug said.

Upstairs in the corporate offices, Chris Finnoff, the founder and president of Turbowest, sat behind a large conference table. "This is Dave Soucie," Doug told him. "The guy I told you about." I was surprised to learn that Doug had mentioned me before. Doug proceeded to embarrass me with compliments, but after the past couple of years, it was music to my ears.

Before I left the office, I accepted the job as night-shift manager of the maintenance department at Turbowest, and just like that, I was back in aviation.

CHAPTER NINE

Hooter and Bagel Boy

Unlike Omniflight Chicago, I enjoyed working at Turbowest. There were problems, of course. The maintenance shop, at first, made very little profit and existed primarily to support the sales department. I began hard-selling modifications and upgrades to customers, and the shop quickly began to exceed the profits of sales. With success came a promotion to director of maintenance. By summer's end, my career and my family's life were back on track.

Friday, September 15, 1989, was a beautiful fall day, sunny and warm. Two days earlier, Jill and I had notched our ninth wedding anniversary, but we had postponed our celebration until that night. The chatter at work Friday morning was about the $6.75 billon purchase of United Airlines by its pilots and management officials, a group that included Denver oilman Marvin Davis. In every way, it began as nothing more than a normal workday.

A Piper Cheyenne IIXL, a corporate airplane from Colorado Springs, was in the shop for routine maintenance on its twin Pratt & Whitney Canada PT6 engines. Around midmorning, I told Hooter, a master at rebuilding turbine engines, "I better cancel the test flight. Doesn't look like you'll get these engines done."

Hooter's name was Keith Alan Crooks. He was tall and thin, with fiery red hair and a personality to match. His nickname, *Hooter,* came from the size twelve, banana-yellow tennis shoes he wore, as well as from rumors that his feet weren't his only oversized body part. The thirty-three-year-old mechanic enjoyed bantering with me and scoffed at my challenge. "The Piper'll be ready right after lunch, just like I said. So go ahead, Dave, pack your barf bag."

I laughed, but if Hooter said the plane would be ready, I knew I should schedule a flight. So I did. Once we serviced a plane, I usually joined Dennis, a fine mechanic as well as a pilot, on the test flight. Sometimes Hooter went along, especially if he had done the engine work.

Late Friday morning, Jill phoned and asked me to come home for lunch and watch Tyler while she ran an errand. She was owed a paycheck and wanted to pick it up.

"I can't," I said. "I have a test flight at 2:00 PM."

"I'll be back before then. I promise."

We lived close to the Jefferson County Airport, and with Jill's assurances, I went home for lunch and played with Tyler while I waited for her to return. I watched the clock reach 1:30 PM. Anxiety began to build. It was 1:40 PM. and Jill still wasn't back. Then 1:45 PM. Our house was fifteen minutes from Turbowest; I had to leave right then if I was going to make the flight, but I couldn't abandon our three-year-old. It was nearly 2:00 PM when Jill showed up.

Angry, I rushed out the door, jumped in the car, and raced toward the airport. I kept telling myself I wasn't all that late, and Dennis surely would wait for me. Besides, someone had to go up with him to monitor and check off each piece of equipment as it was successfully tested. I liked to do it myself to show confidence in our service.

As I neared the airport, I didn't see any planes overhead. *Good,* I thought, *they waited.* I began to relax. Then my gaze dropped to the horizon. The gray plumes of smoke I saw shouldn't have been there. My stomach knotted up. Somehow, I knew instantly that the Piper Cheyenne had gone up without me and a tragedy had occurred.

After forty-five years of flying, Donnell Severts had logged more than twenty-six thousand hours, much of it as a corporate pilot.

He currently flew for Widefield Homes, a housing development in Colorado Springs. On Friday, Don, as he was called, came to pick up the Piper Cheyenne. Typically, he liked to hang out and joke with the Turbowest mechanics, but for some reason, he was in a hurry that day. "Where's Soucie? I want to get out of here!" Don said, his voice echoing through the hangar. The sixty-six-year-old still projected the commanding presence of the Navy pilot who had led WWII bombing runs over Europe. "Dennis, you gonna fly this thing, or do I have to do it?" he said.

"Not without Dave, I won't," Dennis told him. "We'll do the test flight, so just wait. He'll be here."

Various people knew I had gone home for lunch, but nobody knew why I was late or that I was on my way. I hadn't taken time to phone the shop before I stormed out of the house, and in 1989, cell phones were rare.

"I'm not waiting," Don insisted. "I'll do it myself."

Hooter was standing by, having already helped Don get the Piper onto the ramp, ready for the test flight. Dennis wouldn't budge, and finally, when Don insisted on going up right then, Hooter agreed to fly with him and do my job.

"Anybody else wanna go?" Hooter asked.

Hooter took the copilot seat and began recording the start temperatures and oil pressures as the Piper Cheyenne IIXL taxied out, raced down the runway, and shot into the sky. Don climbed to 20,000 feet, made a loop around the foothills west of Denver, and, with everything working smoothly, began his approach to land. The entire test flight took only fifteen minutes. "This is Piper November six three X-ray Lima on approach," he told the tower.

"X-ray Lima, you are encroaching the Cessna one five zero at twelve o'clock," the air traffic controller responded. Compared to the Piper Cheyenne IIXL, one of the fastest corporate turboprop airplanes around, the Cessna was a flying Volkswagen Bug. "Please slow your airspeed to 120 knots," the controller said.

Don acknowledged the instructions, pulled back the throttles, put down the landing gear, and deployed the flaps. But slowing a flying Porsche isn't easy when you are on a seven-degree, downhill glide slope. "X-ray Lima, you are still gaining on the aircraft in front of you," the tower reported. "Abort your approach and go around."

To go around meant that Don had to veer away and circle for twenty minutes before beginning his landing approach again. Don didn't have the patience for it, not on this day. "You want me to slow down for that sucker, I'll slow down," he said. Nobody is certain what Don did next, but he slowed the twin-engine Piper until the plane lost its lift and stalled. The nose of the airplane dropped violently and rocketed toward the ground.

One witness said the Piper Cheyenne dropped like "a big rock coming out of the sky." Another witness, a student at Colorado Aero Tech, watched it come in at a ninety-degree angle and crash before sliding into a gully and hitting an embankment. By the time the Piper Cheyenne stopped, only the tail section remained intact.

When I reached the hangar, Mike Turner, a mechanic's apprentice, was the first to meet me. "They didn't wait, Dave! Dennis told 'em to.

The *Daily Camera*, Boulder, Colorado. David P. Gikley, photographer

Said he wouldn't fly the plane, but the old guy, he took it up, him and Hooter."

"Hooter?" I said in shock. "Oh god, no!"

FAA and NTSB investigators quickly arrived on site. From the top of a hill, everyone at Turbowest watched the police control the roads while investigators, firemen, and EMTs worked the crash. Intense flames sent a pillar of black smoke straight up. The sky rained ashes and soot. Slowly, one by one, we drifted back to the hangar where Doug Stimson made the announcement everyone was dreading: "They say no survivors. Don and Hooter didn't make it."

It was during the announcement that I realized I hadn't seen Randy McMurdy—not on the hillside, not here in the hangar. Jack Randall McMurdy, a twenty-five-year-old with a wife and two kids, was a recent Colorado Aero Tech graduate. Turbowest was his first mechanic's job, and Hooter had made Randy his protégé as soon as he arrived. While Hooter was outrageous, Randy was consistent. Every day at the same lunchtime, Randy ate an onion bagel with cream cheese. Every day, he sat in the same chair in the break room, drinking coffee from his "World's Greatest Son" mug to wash down the bagel. That's how, over time, he became Bagel Boy to me. I figured he had heard about Hooter and was off somewhere, feeling miserable. I went to find him. As I approached his toolbox, I noticed the familiar brown lunch bag from a deli down the road. This was strange. It was long after Randy's lunch break.

"Oh shit," I said aloud. My stomach knotted up again. "Randy! Where's Randy?" I began moving through the hangar looking for him. "Has anyone seen Randy?" I yelled.

No one responded. Not Randy. Not anyone. Finally, one of the mechanics said, "I haven't seen him since . . ." He didn't finish. He didn't need to.

I ran across the hangar to where Doug was talking on the phone with Greg Feith, the NTSB investigator in charge. "Tell him there's another person on board. I think Randy McMurdy went with them." Doug was very close to Randy, and tears welled up in his eyes as he struggled to tell Greg.

Moments later, unable to wait any longer, Doug and I ran to the company truck and raced to the accident site. After we showed our identification to the police, Greg authorized us to walk out to the crash. The fire had subsided, but the airplane was still smoldering. Strands of shiny, molten aluminum draped the hillside and the creek bed below, and bright orange tags marking the airplane parts flapped in the breeze. Two soot-covered body bags lay among the debris. The winch on a fire truck groaned as the cable tightened its grip on the tail of the airplane and lifted it slightly. The coroner, covered in mud while working beneath the suspended airplane, yelled out, "We got another one here!" For the first time, it hit me. The third body bag could easily have had my name on it.

I was still in shock when Greg Feith came up. He was an experienced investigator. He also resembled the actor Tom Selleck and was a good PR man for the NTSB. "It looks like engine failure, Doug. This was a test flight, wasn't it? Did your guys do a hot section repair on these engines?"

"No way!" I shouted as I stepped toward Greg. "You're not going to put this on Hooter!"

Doug stuck out his arm and stopped me with a clothesline tackle. "Shut up and go to the truck, Dave."

I knew that Doug was trying to save me from myself as Jill often did, but I still stomped off, mumbling obscenities. As I looked back to glare and curse at them, I tripped over a deep scar in the ground and fell flat on my face. The disturbed ground was far from the accident site, but the upturned dirt and broken sagebrush were fresh disturbances. When Greg and Doug rushed over to see if I was all right, they immediately understood what we were looking at. "You found the scatter point," Greg said. I had stumbled over the point of impact, the scatter point.

At 250 knots, about 300 miles per hour, even the slightest contact between an airplane wing and the ground is a kiss of death. In this instance, the left wing cut a gouge no more than ten inches deep. If Don had pulled the airplane out of the dive just ten inches higher, a second sooner, heaven and earth would still be the same.

Several days later, and even more disturbed by the NTSB decision to blame the crash on Hooter's engine work, I visited the site alone. It was an overcast day, and the wind was carrying autumn with it. I wandered around the crash site for what seemed like hours, looking for something, for anything, but not really knowing what.

After my useless search, I rested on a large rock. Around me, the sagebrush was burned and the ground covered with the dark soot. Small pieces of metal and glass remained where the aircraft had been. I studied the pattern made by the debris still embedded in the hillside and beside the creek. The sun peeked out for a moment and highlighted something in the water—something yellow.

I slid down the hill to find the source of the yellow reflection and saw it instantly—Hooter's banana-yellow tennis shoe. I plucked it out of the water and studied it. "Help me. I know you did the job right. But I need proof."

God forbid. I was talking to a tennis shoe. I had gone from bumbling Inspector Clouseau to wacky Maxwell Smart. I felt ridiculous, so I dropped the yellow shoe in the creek and climbed over the ridge where the fire truck had been when it hoisted up the tail of the airplane.

From this vantage, I looked south at the scatter point. Starting there, I envisioned a straight line out, a line traveling at 250 knots. I turned around and looked behind me. I had never been psychic or heard voices, but that day, I heard a voice in my head say, "Start walking." So I did. I walked about a mile. Every few yards, I would stop, turn around, and realign my path. If anything was out there, it had to be in line with the scatter point.

I was ready to give up, when I turned around one last time to realign myself. There it was! Right in front of me was the copilot's "n1 speed indicator." There was dried blood and tissue on it, so I wrapped the part in my jacket and took it back to the creek. As I rinsed it off, shards of glass fell into the water. The indicator needles were the next to go. That's when I saw two dents in the thin aluminum face of the gauge. The dents showed the exact outlines of the two indicator needles—one at 98 percent and the other at 100 percent. These dents were proof that both engines were working at the time of impact. It was proof that Hooter had not caused the crash.

I notified Greg Feith, and because of what I had found, the NTSB changed its position on the cause of the accident. Hooter was cleared. I couldn't change the tragedy, but solving the mystery and restoring Hooter's reputation still left me feeling exuberant. It also made me realize that I had a talent for accident investigation. It was a startling realization.

I wondered: *is this why I missed the fatal flight?; does life have a plan for me?* No voice boomed out a yes, but I instantly knew the answer. I knew at that moment that I had been twice blessed in life. As a young man, I knew, without a doubt, the woman with whom I wanted to spend my life. And now I knew, without a doubt, the journey my life would take. I wasn't destined for law school or the boardroom of an aviation tycoon. I was meant to make flying safer. I was meant to be an aviation investigator. What I didn't know, however, is that I would have to die before I would get there.

CHAPTER TEN

Dying Lessons

After the deaths of Hooter and Bagel Boy, the usual bantering and laughter that filled the hangar disappeared. Life had failed us. However we explain it, most of us expect life to make sense: for good to be rewarded and bad punished and for competency to keep us safe and foolishness to trigger disaster. But competency had crashed, and goodness lay zippered inside a body bag. Our unease with life's fickleness remained even after the airplane parts, body bags, and orange tags had disappeared, and we quietly did our jobs and waited for a sense of ease to return.

That finally happened a couple of weeks after the crash. It happened the day I saw Walt Wise running into our hangar yelling my name. "You gotta hide me, Dave!" he said as soon as he saw me.

"What're you talking about?" I said.

"The airport cops are chasing me." Suddenly, I saw the flashing lights of police cars headed toward the hangar. Walt saw them too. "I was at the flight school, and they said you worked here, so"—he paused and grinned at me—"so I ran across an active runway."

Without asking him why, I hustled Walt into the parts closet of the engine shop, while the police cars rolled on past. During this time, I heard a muffled sound in the hangar. I walked toward it and saw fifteen grown men laughing. I joined in. Thanks to Walt's Keystone Kops caper, ease had returned. It had been two years since I'd seen him, and when I finally stopped laughing, I said, "What was so important that you'd do something like that?"

"Dave, the FAA is hiring." Walt handed me a glossy brochure. "Federal Aviation Administration" boldly encircled a golden flaming torch on the cover. "The FAA really needs someone like you, Dave," he said. "They need your heart. They need people who truly care."

Walt had nearly gotten arrested in his eagerness to tell me the FAA was hiring. Despite this, the brochure stayed in my desk drawer for weeks. Every time I reached for a pen, I saw it. Finally, one day, I picked it up and read it: "Aviation Safety Inspectors (ASIs) are the FAA's on-site detectives." That sounded interesting, so I read on. "Inspectors develop, administer, investigate, and enforce safety regulations and standards." This really got my attention. "Oversee the safe production, operation, maintenance, and modification of all aircraft flying today." Safety? Risks and hazards? This was exactly what I had been looking for all along. That day, I completed the application and sent it in. Walt was excited when I told him, but when I learned that "the FAA is hiring" didn't mean there were current openings, it dampened my enthusiasm some. "Just wait, Dave," he said. So I did.

A few months later, Turbowest began cutbacks. For years, an oil boom had provided Colorado with a second gold rush, but like the first one, it, too, had dried up. At the same time, new tax laws made buying corporate airplanes less appealing. Lean times meant cutting jobs. The word *downsizing* was not part of everyday vernacular in 1990. When Gene Langefeldt, the CFO, decided to cut the employees with the highest salaries, my name was near the top.

I was let go from Turbowest early in the afternoon, but instead of going home, I drove around aimlessly. I couldn't face being unemployed again. I couldn't face telling Jill. Hours later, I parked the car and watched the sun setting over Long's Peak. My perch also overlooked Jeffco Airport. While sitting there, I noticed the sign for Colorado Aero Tech across the street. "I wonder if they're hiring teachers," I said aloud. I wondered. I started the car and raced to the school parking lot.

The first person I saw was Sherry, a sweet woman from Oklahoma. Ten years ago, when I had walked through the Colorado Aero Tech doors as a student, she worked the front desk. She still had the same large, coiled bun of hair and strong scent of perfume, but now she had her own office and was in charge of personnel. "Well, hi, David," she said with a twang. "Haven't seen you for a while." She loved it when former students dropped by to visit.

I dodged her questions about how I was doing by saying, "You're sure working late today."

"Well, I need to find a good instructor for the aircraft-systems class. Think I could convince you?"

I took the job out of desperation while I waited for the FAA. As the months passed, I found, to my surprise, that I loved teaching. Even more surprising, the students responded well to me. Most of the thirty students in my class worked day jobs and started class in midafternoon. Most of them arrived tired. At first, I resorted to cheap tricks to combat their nodding heads. I dropped large aircraft wheels on the floor. I told grand stories of my adventures in aviation. With my arms outstretched, I glided through the classroom, describing the Bernoulli principle of flight. My cutting torch melted various parts of their desks to show how molten aluminum differs from molten steel. I jumped onto a table and, spinning in circles and beeping, demonstrated the difference between unidirectional and omnidirectional ADF antennas. I was part teacher and part entertainer. It was fun. *If this is my future*, I thought, *it wouldn't be so bad.*

Summer slipped into fall without a word from the FAA. "I'll be back before the thirteenth," Jill told me one day while loading a suitcase into her brother's car. I mumbled something. "You forgot it, didn't you?" she said.

"Of course not. I know it's our anniversary." In truth, I often forgot our anniversary, and she knew it.

"It's our tenth, Dave, our tenth anniversary." She settled in the driver's seat and started the engine. "Remember what you promised me?"

I could still lie when necessary. "I remember." I gave her another kiss and closed the door. "Go have fun."

That night, I tried to imagine what promise I had made. At one point, straining my mind to remember, I was twisting my wedding band. Then I remembered—I had promised Jill a nice diamond ring on our tenth anniversary. Oh crap!

Not only had I not saved any money for the ring, but we could barely pay the bills. I tried to think of an alternate gift, something that would impress her without costing thousands. It needed to be something she really wanted so she would forgive me. The only thing I came up with was the bedroom. Jill really wanted us to redecorate our bedroom. Could I afford it? What if I did it myself?

The trip to Florida would take five days. Brad, Jill's brother, had enrolled at Florida Institute of Technology, and Jill had agreed to drive him there and fly home. Five days, two jobs, no money. Piece of cake!

I bought a beautiful bedspread, picked out her favorite paint colors, and found window coverings that I knew she would like. Tyler was my only helper. At four years old, he really enjoyed painting the walls, and the carpet, and the new bedspread. After this bonding experience, I finished the painting when Tyler was asleep.

The day before Jill was to return, the job was nearly complete. After putting Tyler to bed, I worked until the wee hours of the morning. At around 2:30 AM, I dragged a kitchen chair to the window to use as a stepladder to hang curtains. At one point, I couldn't reach far enough, and I was too tired to climb off and move the chair, so I stretched out as far as I could. The chair began to topple. I grabbed on to the new, white curtains as I flipped forward. My legs flew up over my head, like the day I was flopping around in the Learjet 35, but this time, I was not weightless. This time I was falling face first toward the upturned chair legs.

At first, I thought I was simply dreaming; I watched Jill on the airplane flying back from Florida. She was in an exit row and reading a Hollywood gossip magazine. She was smiling. Then I saw my mother in her bed, sleeping peacefully. I kissed her on the cheek. Then I saw Tyler get out of his race-car bed and walk into our bedroom to crawl beside me to snuggle. But my bed was empty. "Daddy, where are you?" he said. That's when he saw me lying on the floor beside the bloody, white curtains. He screamed. Then he ran. He ran through the house, screaming for help, but he was alone. *Enough!* I had to wake up and go to my son. I reached for my neck and found a large piece of wood sticking out of it. After pulling it out, I poked my fingers inside my neck to feel for a pulse. I found it. A pulse meant I was alive. I woke up!

The blood on me was real. The chair leg broken off in my neck was also real. The only good part of separating reality from the dream was the realization that Tyler was still asleep. In this, I had seen the future, not the present, and I had to change it. I couldn't let him find me. I couldn't let him be terrorized by something that would haunt him his whole life. I had to reach a phone.

I couldn't get up, so I crawled. A streak of blood followed my path. It seemed to take forever to get to . . . where was I going? Why was I on the floor? Why was I bleeding? My mind wasn't working any better than my body.

Eventually, I reached a dresser and pulled the phone off and onto the floor. I dialed 911 and tried to tell the operator that my son was alone. "Get him out," I said. The female voice assured me that help was on the way. She told me not to worry. Was I worried? I lay there, listening to her, until I heard the ambulance sirens. Then I passed out. When the paramedics carried me outside, I regained consciousness long enough to see Debbie Pearce, our next-door neighbor, holding Tyler in her arms. He was safe. It was time for the big sleep.

The first thing I saw was a smooth, golden reflection. I couldn't turn my head to see what it was or from where it came. I willed it closer. It came. Amazing! I could change things simply by willing it. Up close, it appeared to be a small column of water shaped like a curious snake slowly assessing my condition. The golden reflection came from its watery skin. In the reflections, I could see people working and going about their business. Some of the reflections moved with ease, in concert with this being. Others moved against the flow. They struggled and toiled. I wanted to see more. I wanted to see what was driving this liquid energy, to see inside the watery reflections. Just as before, the watery creature obliged my wish. It began to flow over me. It was impossible to say whether I became it or it became me. The only physical change I felt was a pleasing happiness. For the first time in my life, I felt the absence of boundaries or limits. All my selfish wants and needs were stripped away. I felt a connection to all that is. I understood that all things exist because of this common flow. I knew that I belonged here, that I . . .

"Whew! He's back!" The nurse in the ambulance put her hand on my forehead and leaned over to look into my eyes. "You here, buddy? You with me? We lost you there."

Later, the emergency-room doctor, a serious-looking, older man, entered the room to examine my chart. "What's the diagnosis, Doc?" My raspy, low-pitched voice surprised me. "Hey, I sound like Kris Kristofferson. 'Freedom's just another word . . .'"

"Save it," the doctor barked. It was clear he didn't appreciate my humor—or my singing. "You're sure one lucky son of a bitch," he said. "Another fraction of an inch and that chair leg would've taken out your artery. You'd have been dead in seconds." He stood in disbelief, looking at me as if I was an alien specimen. "Your heart stopped for nearly three minutes, Mr. Soucie. You're lucky to be alive."

I laughed so hard I nearly busted the stitches in my neck. "No, Doc," I said. "I'm lucky to have died."

He looked at me as though I were crazy. I knew it would be useless to try to explain to him that dying had let me see my life clearly, that it had shown me a man consumed with self-pity and anger ever since being axed at Air Methods. I had forgotten about my dreams. I had been obsessed with showing Roy Morgan that he had made a mistake, but the real mistake was the one I had been making over and over—the mistake of looking for a predictable future. We try so hard, out of fear, to avoid change; yet change is the only certainty life offers.

Miraculously, I left the hospital later that morning, the day Jill was due to return. The telephone rang as soon as I walked through the door. The handset still had blood on it from the night before. My voice was scratchy when I answered and said hello.

"Is this David Soucie?"

"Yes," I croaked. "Who's this?"

"This is Peter Undem."

I don't know a Peter Undem, do I?

He must have heard my thought. "I'm with the FAA in Honolulu," he explained. "I'm calling to offer you a position as an inspector."

CHAPTER ELEVEN

Bloodhounds and Bookkeepers

After a wave of glee subsided, I asked Peter Undem why someone in Hawaii was calling me about a job in Denver.

"David, the FAA position that's open is in Honolulu," Peter explained. "Is that a problem?"

This news should have set off many alarms—our son's schooling, Jill's job, the sale of the house, the higher cost of living, the change from mountain foothills to a tropical island, and a dozen other concerns—but I didn't hesitate. "No," I said. "Not a problem at all."

A short time later, I left to pick up Jill. Upon her return, she was expecting a diamond ring for our tenth anniversary, not a major life change. I didn't know how to tell her that the man she was seeing today was not the same man she had kissed and said good-bye to five days earlier.

At the Denver airport, I greeted Jill with a kiss and an abnormally passionate hug. She didn't say anything, but she did give me a questioning look. We were in the car, and I was driving away when she noticed the bandage. "My god, David, what happened to you?" She gently reattached the tape that had come loose during our embrace.

Tears flooded my eyes until my vision blurred. I pulled off onto the side of the road as cars zipped by. With the whine of rubber serenading us, I told Jill everything—from my attempt to redecorate our bedroom to seeing the golden serpent to the FAA's phone call. She sat quietly, listening and all the while looking out the window. Finally, I confessed, "I did what I did because there isn't a diamond ring as I had promised you."

"I know that," she said. Only then did she turn to me and smile. "David, I've had a front-row seat. I've watched you struggle with your conscience since Mike's death, and I couldn't change it. I know you feel responsible and need to make it right." As usual, Jill seemed to know me better than I knew myself. "I'm just happy you've finally found a way."

My employment interview was scheduled for the following afternoon at the Denver Flight Standards District Office. Unlike my prior experience with government agencies, the FAA was moving quickly. I didn't stop to wonder why. I just assumed they really wanted me.

The interviewer was behind a desk when I entered his office. He looked familiar, but I couldn't quite place him until he greeted me. Gary Gomes had added a little weight and a lot of gray hair since we had met while he was investigating Mike Meyer's fatal accident.

We chitchatted without either of us mentioning the accident or even having met before. Then Gary asked a few routine questions, checked a few boxes, and said, "That's it." He made an official job offer and told me the position in Honolulu paid twenty-five thousand dollars a year, plus a fifty-two-hundred-dollar relocation allowance. "Do you accept the job offer, Dave?"

I nodded yes and said, "I do." It felt like marriage.

"Good, then we'll make it official and swear you in."

"Now?" At the least, I was imagining Jill by my side and a rubber-chicken dinner with toasts and speeches.

"Yeah. You have to be in Hawaii by October first." Without pausing, Gary told me to put one hand on the Bible and the other on my heart. "Do you solemnly swear to uphold the Constitution of the United States of America?"

Wow! The impact of the oath made my heart pound. I always felt I had neglected a responsibility by not serving in the military. This felt like a second chance. I said "I do" and then repeated several vows and ended with another "I do." Only then did I fully realize what Gary had said. "When you said I'd be in Hawaii by October first, you meant Oklahoma City, right? For my training?" Walt had told me to expect a few weeks of training at the FAA Academy. "That's pretty fast, twelve days away."

Gary went rigid in his chair. "No, I said Hawaii, and I meant Hawaii. You *will* be there October first."

This was my first experience with a direct order, and I didn't like it much. I started explaining why we needed more time.

Before I got halfway through my list, Gary said, "It's October first or no job." Maybe I looked stunned, or maybe he remembered my passion for improving EMS helicopter safety, or maybe he was just a nice guy. "It's like this," he said in a softer voice. "You're there by October first, or the FAA loses funding for the position. It's about budget deadlines. That's the reason you got hired now. Use it or lose it. That's life in the government, Dave."

"So when do I go to the FAA Academy for training?"

"Later. The academy's booked up for at least a year."

I heard him, but my brain was rebelling. The FAA was actually going to send me out to work as an FAA safety inspector without any training whatsoever? Surely I was missing something. Not knowing what to say, I sat quietly.

"Don't worry," Gary said. "They'll keep you busy in Hawaii." A moment later, he left to get some paperwork. I sat there and tried to imagine what "keeping busy" in the FAA meant. Wham! A large three-ring binder was slammed onto the desk in front of me. I read the title.

The Federal Aviation Administration
Providing the World's Safest Aerospace System
Inspector Familiarization Manual

"This is your training until the academy opens up," Gary explained. "It's all we can offer with our limited resources." Gary picked up his overcoat, shook my hand, and escorted me to the door. "Good luck, Dave, and aloha!"

When I told Jill the details of the job, her first reaction, like mine, was shock. Afterward, we spent hours examining our finances and every what-if we could imagine. By the time Jill gathered up the bills and bank statements and attempted budgets from the kitchen table, we both felt a sense of defeat. She laid her head on my shoulder and whispered in my ear, "It's just not possible, David. You'll have to go without us."

"What if I don't go? I can wait for a job here."

She shook her head no. "No more what-ifs. You go, and I'll stay here with Tyler until you're ready for us."

Five days later, I was on my way to Hawaii.

During the six-hour flight, I was wedged between a huge Samoan man and a young Hawaiian girl with a baby. I had a new portable CD player to listen to and the thick FAA manual to read. I started with the FAA manual.

The first lesson covered the history of the agencies that preceded the FAA. I read about the Air Commerce Act of 1926, which charged the Commerce Department with encouraging civil aviation as well as with both setting and enforcing the standards for both planes and pilots. This part grabbed my attention because of its contradiction. Telling an agency to both "encourage aviation" and "set standards" sounded self-defeating. Do policemen encourage drivers to buy faster cars and drive more? Does the IRS publish *101 Ways to Reduce Your Taxes*?

I read on, certain that this contradiction had changed. But I was wrong. For many years, there were not only dual objectives, but there were also dual agencies. This lasted until 1958, when Congress created the Federal Aviation Agency. The FAA was responsible for making rules, for enforcing them in the safety field, and for maintaining a common civil-military system of air navigation and air traffic control. The agency was renamed the Federal Aviation Administration (FAA) in 1967 and placed within the newly created Department of Transportation. At the same time, accident investigation was transferred to the NTSB. After reading that the NTSB was responsible for accident investigations, I became concerned.

I recalled walking the crash site at Turbowest and feeling thrilled when I cleared Hooter of any blame and proved the investigation wrong. That was the reason I had applied to the FAA—to investigate accidents. I wondered if I should have applied to the NTSB instead.

I later discovered the NTSB frequently has the FAA handle its accident investigations for them, and even if they don't, the FAA is always involved.

The baby next to me began crying. I recognized the sound. The baby was hungry.

"He's hungry," I said to the Hawaiian girl. "My son would cry the same way when he wanted to be fed." The girl gave a slight nod and began to feed the baby. The cries died down instantly. Ten minutes later, the baby was asleep. I missed Tyler and Jill already. The mother said something to me, a Hawaiian word. I didn't know the word, but I knew from her eyes it meant thank you.

I returned to reading, and much of what I found, I already knew: the FAA has three divisions—Airports and Facilities, Air Traffic Control, and Aviation Safety. Aviation Safety was what interested me. It is divided into Aircraft Certification and Flight Standards. Aircraft Certification oversees and regulates the design and manufacturing of all aircraft, while Flight Standards oversees and regulates the certification of airlines, pilots, and mechanics, as well as the certification and operation of all aircraft. My job was in Flight Standards. (For more information on the FAA, see appendix.)

The next section covered accident investigation, and it was both fascinating and perplexing. The manual explained that, while the FAA investigates accidents, the NTSB is exclusively responsible for determining the root cause of an accident. The FAA, on the other hand, investigates an accident to determine whether or not any of its nine functions were involved as causes of the accident. I studied the list. (See appendix for the complete list.) It seemed strange that the first-listed FAA responsibility (function) was to investigate an accident to determine if FAA functions were a factor in causing the accident. Equally perplexing was the fifth function on the list. The investigation considers whether the rules written by the FAA (CFR 14) are adequate to determine if the rules it has written are adequate for the investigation. Huh?

An overhead *ding* sounded. I suddenly got it. The FAA investigates accidents to look for someone or something to blame. *Ding!* Enough.

I got it! Actually, the *ding* was to announce our descent into Honolulu. Just in time. I felt as though I had already reduced my IQ by twenty points.

The FAA administrative officer picked me up at the airport. She was a small Polynesian woman. She told me her name, but it had an equal number of vowels and consonants, and my tongue couldn't keep up with my lips when I tried to repeat it. After I made several mangled attempts to pronounce it, she told me to call her Dee. Dee took me into her office to complete the paperwork and to take a photo for an official ID. The rest of the day was spent meeting most of the thirty inspectors and support staff who worked at the Hawaii Flight Standards District Office in Honolulu. I was the only man wearing a suit. The entire group welcomed me with gifts and advice and food. Lots of food! It was Aloha Friday, and everyone had brought a tasty local dish for the weekly celebration. It was a Hawaiian version of a potluck. They called the food "pupus."

On Monday, Dee brought me a box of her husband's old Hawaiian shirts and told me to ditch the suit. Like Alice, I had slid down a rabbit hole into a strange and wonderful world. I couldn't wait for Jill and Tyler to join me.

I rented a room in the lower level of a large, old house on the Kelanianioli Highway. Tommy, a hippy entrepreneur who ran Tommy's Tours and who loved to show people the islands, owned the house. The house needed repairs, so I traded construction work for my rent. It also was right on the beach. When the trade winds blew, the wood house moaned and creaked like a huge sailing vessel, while ocean waves pounded rhythmically outside my door.

I spent my days in a "nice" government-issue office, reading and organizing regulations and handbooks.

Part of my job was to catalog pilot records and to administer tests for airframe and powerplant mechanics, the same test I had taken years earlier. I quickly grew tired of the routine, so I went out on inspections with veteran inspectors Scott Christiansen and Les Sergeant any chance I could. Almost from the beginning, I begged Peter Undem for more interesting things to do. Peter was the manager of the Air Worthiness division, one of the three divisions in the Hawaii FAA office. The other

My FAA office in Hawaii and me with a *Magnum, P.I.* mustache

two divisions were Operations and General Aviation. Since I worked for Air Worthiness, Peter was my direct boss.

One day, he rolled a cart that held about twenty three-ring binders and several file folders into my office. "This is the file for the Aloha Airlines' Flight 243 accident," Peter said. "I want you to review the files thoroughly and complete an analysis report."

"Analysis of what exactly?"

"How well the FAA did its job when we investigated."

Once Peter left, I stared at the huge binders, stack after stack of them. *Well,* I thought, *you asked for it.* Maybe my new job was turning out to be more bookkeeper than bloodhound, but I relished the challenge and dug in.

CHAPTER TWELVE

The Bambi Hazard

A newspaper photo of a Boeing 737 lay on top of the binders and file folders stacked on the metal cart. The top of the airplane looked like a sardine can peeled open with a key. I remembered the photo and the news coverage of the disaster a year earlier. At the time, I had no idea that I would be digging through the accident reports like an archeologist in search of hidden truths. I read and reread a summary of the disaster:

> On April 28, 1988, the aircraft, *Queen Liliuokalani*, a Boeing 737 owned by Aloha Airlines, took off from Hilo International Airport at 1:25 PM bound for Honolulu. Ninety passengers and five crew were on board. No problems were reported during take-off or ascent.
>
> Thirteen minutes after take-off, when the aircraft reached its normal flight altitude of 24,000 feet (7,300 m), a small section on the left side of the roof ruptured. This caused an explosive decompression, which tore off the entire top half of the aircraft roof from just behind the cockpit to the forewing area.
>
> Chief flight attendant, Clarabelle "C.B." Lansing, was standing at row 5 collecting drink cups from passengers. At the moment of

decompression, she was ejected through a hole in the side of the airplane.

The word "ejected" bothered me. Movie scenes of someone being sucked out of a depressurized airplane flashed before my eyes.

In the cockpit, Captain Robert Schornstheimer looked back and saw blue sky where the first class cabin's roof had been. He immediately contacted Kahului Airport on Maui and declared an emergency.

The explosive decompression severed the electrical wiring from the nose-gear to the indicator light on the instrument panel. Due to this, the pilots had no way of knowing if the landing gear was fully lowered. The passengers were told to don their lifejackets, in case the aircraft didn't reach land.

At 1:58 PM the airplane touched down on Kahului Airport's runway 2. Upon landing, the passengers who were able to do so quickly escaped the broken plane by using the aircraft's emergency evacuation slides.

Associated Press library photo • April 28, 1988

The island only had a couple of ambulances, so the injured were taken to the hospital in tour vans provided by nearby Akamai Tours. The safe landing of the aircraft with such a major loss of integrity was unprecedented.

Unprecedented? By all rhyme and reason, everybody on board should have died, yet there was only one fatality. How was I to go about dissecting a miracle? I had no answer, so I lay the photo on the credenza behind my desk and picked up the top file. It was the NTSB report, a good place to start digging. Once I did, the keywords section on the second page of the NTSB accident report stunned me.

17. KEYWORDS

Decompression; Disbonding; Fatigue Cracking; Corrosion; Multiple site damage; FAA Surveillance; Maintenance Program; Non Destructive testing.

I had read hundreds of NTSB accident reports before. It was highly unusual to see a report for a single accident that had so many significant causes in the keywords section.

The NTSB report also detailed a string of other safety oversight failures. The FAA inspectors assigned to Aloha Airlines should have done a better job overseeing maintenance at the airline. Aloha mechanics should have done a better job of inspecting the aircraft for fatigue cracks and corrosion. (Later, I discovered that the airline mechanics did find cracks and had reported them to both Aloha management and Boeing engineers, both of whom had failed to act.) The airplane, more than nineteen years old, should have been taken out of service by the airline management because, after some seventy-five thousand flights, the lap joints that held the aircraft together were corroded. This problem was previously noted on inspection reports, but the management had again failed to take action.

The NTSB skewered both Aloha Airlines' management and Boeing executives throughout the report. When I finished reading the report, I was confident that each of these causes factored in the accident, but I didn't see any of them as the root cause. I was sure the root cause was buried deeper.

At about ten o'clock that night, Steve Badger, the manager of the operations section, wandered in to see what I was doing there so late. (It didn't occur to me to wonder why he was there.) "Hey, Gadget, working late?" he said.

I had developed a program to search the old Burroughs' mainframe computers in the office for information about the surveillance activities of the inspectors. For this, I was nicknamed Inspector Gadget. "Steve, can I ask you a question?"

"I think you already did, Gadget." Steve could never resist a joke, no matter how lame. It was the price I paid for his help. I was concerned that I had missed something in the report. After I told him my concerns, he said, "Let me tell you a story, Dave." He pulled up a government-issue chair, leaned it back at an angle, and plopped his boots on my desk. "Imagine driving down the highway minding your own business. As you come around a blind curve, a deer jumps into the road directly in front of you. You slam on your brakes but can't stop. You try to steer around the deer, but the road is wet and your tires are bad and you slide out of control. Your valiant efforts to save Bambi's life end in a loud *thunk!* Bambi's dead, and your car is wrecked."

"What the hell are you talking about?"

"Think about it, Dave. How would the NTSB write the Bambi accident report?" I knew it was a rhetorical question. "Well, I'll tell you how it'd go." He began to speak as if reading from a report: "The owner of the vehicle should have overseen the maintenance better, including replacing the tires and brakes more often. The driver should have been going slower so he could have seen the deer sooner and avoided the collision." By now, I was nodding my head yes. "There should have been a warning sign that deer crossed the road at this bend. If there had been, the driver would have known to be more cautious."

"I get it, Steve," I said.

"Just wait, Dave. Here's the point. Do you think the NTSB report would have asked why a fence hadn't been built to prevent the deer from crossing the highway at the blind curve?" I shook my head no. "Understand the difference between a diagnosis and a prognosis, Dave.

A diagnosis analyzes why the driver didn't stop the car before killing the deer. A prognosis says that deer will continue to be killed until we build a fence and remove them from the equation." The boots came down, the chair legs landed with a *thump*, and Steve sprang up and grabbed his umbrella. "Gotta go home, Gadget."

I sat back and processed the Aloha Airlines' accident as if it were the deer story. The FAA and NTSB diagnose aircraft accidents. They analyze the reasons they happen. By doing so, they can assign blame. That's why the NTSB needed a grocery list of causes. It was their way of listing all the things the driver could and should have done. The prognosis—fence building—was nowhere in the report. The prognosis for avoiding this accident was simple: remove the fatigue cracks from the equation.

The Boeing 737 was designed to withstand a tear in the skin and rapid decompression, but in this case, multiple cracks in the fuselage all failed at the same time when a seam tore loose. But what caused the cracks? Failing to answer that was like saying, "The deer just jumped out of nowhere; nothing I could do about it." Case closed. I combed the report until I found the place where "multiple site damage" was noted as a factor in the accident. But even here, there was nothing to address what caused the metal fatigue in the first place. I needed to find the one factor that, if removed, would have prevented C. B. Lansing's death. I needed to remove Bambi from the equation.

At least I knew what to look for now, so I scanned the report again. And I found it! Section 1.6.1, Aircraft General, listed the flight hours of the Boeing 737—35,496 hours during 89,680 flights. A simple math calculation led me to discover the root cause of the accident. It was not obvious, for it had occurred nearly thirty years earlier at the Boeing factory in Tacoma, Washington.

In 1966, a young aircraft engineer made an assumption that the small Boeing 737 would be used for flights lasting two hours or longer. It was a reasonable assumption at the time. Pressurized jets were used only for longer trips, not for hopping from island to island. This usage meant the pressure on the fuselage would expand and contract no more than once every two hours. Given the flight hours and the number of flights listed, the average flight time was about twenty-four minutes.

In reality, for the past nineteen years, the metal had expanded and contracted every twenty-four minutes, not every two hours. This was not something the engineers had considered when they designed the airplane, although they had put a limit of seventy-five thousand cycles on the airframe. Above that number, the engineers warned, the metal airframe would be work-hardened and fatigued, resulting in cracks and explosive decompression. Cracks and decompression? This sounded horribly familiar.

Even after considering the dated and erroneous engineering assumption, I still couldn't understand why nobody prevented the disaster. There were plenty of other clues to suggest that the risk level was high. So why didn't somebody act on it? Who was really at fault?

By the time I closed the last binder, the night janitor had finished cleaning and had turned out all the lights. In the dark, the green exit sign over the door in the hall shone with an eerie glow. As I pushed the cart toward the library, the wheels wobbled under the weight of the files, and the casters squeaked for lack of oil. Churning in my mind were angry thoughts about callousness and carelessness and greed. I wanted to blame somebody. But who? The aviation industry? The FAA?

"It's you!" a voice said from somewhere down the hall. The noisy cart squeaked to a stop as I clenched the handle and froze in fear. I strained to hear the voice again but heard nothing. A chill shook my body. I shoved the cart into the library and ran for the door.

Mounakea*, a local who came to the beach to harvest limu seaweed nearly every night, was lying in my hammock and drinking a beer when I got home that night. After I pulled a beer from his cooler, we sat together and watched the waves. "Look like you seen a ghost, bra."

I shuddered. "More like I heard one." I repeated what had happened. Only then did I realize what it meant. I was angry at Aloha Airlines and the Boeing engineers and everyone else because I had done the same thing myself.

Mounakea sat with me late into the night while I told him about Mike Myers and the wire-strike-kit disaster and how the incident had driven me to Hawaii to work for the FAA. I confessed my guilt, admitting that I could have prevented Mike Myers's death as easily as

Aloha and Boeing could have saved the life of C. B. Lansing. It nearly broke me to admit that I was no different from them.

"No, bra! You okay. But you gotta forgive yourself." He put a big, chunky arm around my slumped shoulders. "Ya do da best ya can wit what ya know, bra. Problem is, bra, ya just gotta take da time to get clear 'bout what ya know and what ya gotta do 'bout it." His simple interpretation in island patois felt more accurate to me than a year on a shrink's couch. I watched him pick up his buckets of limu seaweed and walk to the street. He turned and gave me the island greeting shaka sign with his thumb and little finger in the air. "Laidahs, man."

The next day, I wrote my final report reviewing the Aloha accident. I suggested that the right people (i.e., the airline managers and aircraft manufacturing executives—the decision makers) had the information needed to recognize the risk of continuing to fly the aircraft before the accident occurred, but they lacked a method to evaluate the importance of the information. I recommended that three questions concerning safety be asked during any aviation-accident investigation: Was information available that could have prevented the accident? If information was available, did the right people have access to it at the right time? If the right people had the information at the right time, did they recognize it as critical safety information and take action?

I also recommended a standardized method be implemented for decision makers to access, as well as share, information about mission critical operations, routine maintenance, and unscheduled service events. Sharing information among competing airlines was asking a lot. I was certain it was the only way decision makers could make accurate risk assessments about safety. And finally, I noted in the report that failure of critical aircraft components is highly probable without a standard process to continuously, for the entire life of the aircraft, validate operational assumptions made by engineers during design.

Since my assigned job was to evaluate the FAA investigation, I suggested that the nine responsibilities written by the FAA in 1958 were insufficient to address the current complexities of the aviation business. I recommended two additional responsibilities be added by the FAA. One, to determine if information that could have prevented the accident was available prior to the accident, and two, to determine if decision makers had information prior to the accident but failed to recognize it as a precursor to the accident. (See appendix for additional information on all the above.)

When I handed my report to Peter Undem, I felt very good about it. I felt I had done something that would make a difference, and I was excited to hear what Peter would say. He took the report and looked at me curiously before saying, "I had hoped that this would take you longer, Dave."

I wondered what he meant. Was the job assignment meant to be busy work and nothing more? I left his office feeling greatly disappointed.

Weeks passed before Peter read the report and told me I had done a good job. A good job? I was expecting something like, "Oh my god, Dave, we better change the way we review accidents." Good job? Wow!

Despite Peter's lackluster response, once my life returned to the routine of watching mechanics take tests and shadowing FAA inspectors on mundane inspection trips, I soon longed for another assignment. Anything! I never once imagined that when my new assignment came, it would involve carrying bombs and guns through the airport.

CHAPTER THIRTEEN

Human Sacrifice

The day after Thanksgiving, Gary K., a veteran inspector, took me to the airport with him to pick up Pete Beckner, the manager of our office. Gary's last name was something like Kamea'hemea, so he was Gary K. to nonlocals. Pete's plane was late, so Gary and I found an airport café where we could enjoy the Hawaiian weather and some Kona coffee. With the scent of flowers and the sounds of vacationers filling the air, we didn't mind waiting. As we people-watched, I noticed a security guard creeping up behind Gary. He put a finger to his lips to silence me. A moment later, he smacked Gary's back. "Howzit, bra."

They engaged in a ritualistic greeting of colliding hands and mashing knuckles. Gary responded, "Hey, cuz. Howzit, brudah? I hear you run da show now, brudah."

Cuz, brudah, and wazzit? I longed for Barbara Billingsley from *Airplane!* to jump up from her seat: "Oh, Inspector, excuse me, I speak jive." Even without her help, I figured out that Gary's "cuz" was not actually related, but that they had once worked together, and, most importantly, that the guard was in charge of security for one of the airlines. In 1990, the airlines managed terminal security and employed the screeners, while the FAA was responsible for investigating how well they did the job. Once Gary and his cuz finished with their extended reunion, the guard looked at me and said, "You ready, bra?"

"Ready for what?" I asked.

"He don't know?" said the guard. Gary shook his head. They both laughed. The guard said he wanted my help and asked me to go with him. I got up, but Gary didn't.

"I'll be right here when you get back, Dave," he said.

After we left Gary, the guard explained how he wanted me to help. He said that airport security liked to use new FAA inspectors to test the security checkpoints. "You got a new face, bra. Das why we need you." He laughed again. "Ever'body get to do it, bra."

The guard led me to a small, dark room where nameless men gave me a ten-minute briefing. The men in suits were probably FBI. One scruffy guy was likely DEA. Others were FAA security officials. It wasn't a friendly bunch, but I hesitantly agreed to their request anyway. In today's world, it's hard to imagine anyone giving an FAA rookie a briefcase filled with guns and a bomb and sending him through security, but that's what was about to happen.

The briefing finished, and everybody stood up. By now, any fleeting delight in playing James Bond or Eliot Ness was overshadowed by my Walter Mitty heart beating so loudly that it competed with the clickety-clack of the overhead fan. "Don't look back at us," one of the suits hissed at me as he opened the door. "We'll be right behind you."

As I walked off, I tried to calm down. I kept telling myself that I could do this, that I could trust Gary, that he wouldn't let me get into anything too bad, and anyway, the suited men were there behind me. I peeked back.

"Turn around!"

"Oh shit, I forgot."

"Shut up!"

"Okay." When I'm nervous, I have to get in the last word. I forced myself to keep quiet and to keep walking. A hundred feet before the checkpoint, I heard the footsteps behind me suddenly stop. I resisted another urge to look back. The briefcase in my hand felt as though it weighed a ton.

Beads of sweat collected on my face as I set the briefcase on the belt. I darted through the metal detector, relieved that the small handgun in my jacket didn't set off the alarms. Whew! In 1990, the checkpoint sensors measured only the amount of time a mass of metal was in front of them. If you went through quickly or turned sideways, it wasn't hard to get a gun through the security sensors.

Even so, my sprint through the sensors caught the attention of the young, female security guard. She looked over at me and then at the X-ray screen as my briefcase went through; her face literally turned white from blood loss. She had seen the translucent images of all the weapons. *Good*, I thought, *the game's over.* I smiled and winked at her. Her response was to push the silent alarm.

Policemen appeared as magically as rabbits from a hat. They pulled my arms behind my back and slammed my face into the moving rubber belt. I managed to lift my head to look for the agents who had followed me. They were gone. "Shit! Where are you assholes?" I shouted.

The young woman stepped out from behind the scanning machine with the briefcase in her hand. "What have we here?" The handcuffs dug into my wrists as a policeman used them to force my face back down on the rubber belt. "You're not winking at me now, are you?" the woman said, gloating. She flicked the two locks and opened the case.

"No! Don't!" yelled the policeman. He jumped behind me to shield himself from the explosion. Of course, the bomb wasn't real—just an alarm clock and some wires and Silly Putty or such. If the bomb had been real, the young woman would have blown us all to bits.

Only then did the black suits and the scruffy one and the others suddenly reappear. Within seconds, I was out of the cuffs. I rubbed my hands while a crowd of onlookers looked at me like cannibals visiting a fat farm. The agents formed a wall around us. Right before one of them dropped a paper bag over my head, I saw another agent fire the security guard because she had screwed up and opened the briefcase. Then I was escorted away. Moments later, the agents debriefed me in the same small, dark room with the thumping fan. It didn't take long. Mostly, they warned me not to talk about what had happened, and then they thanked me for my service, and we all left.

Once I had collected myself, which took a while, I realized that many things about the security test bothered me, especially the realization that the young woman was a scapegoat. No doubt the various reports would say the system worked fine, but that security personnel failed to follow procedures. In truth, nobody was actually looking at how well the system worked. The airport-security exercise was dramatic proof of the conclusion I had first reached on the airplane to Hawaii while reading the FAA Training Manual. After being baffled as to why the FAA even investigates aviation accidents, if not to look for the root cause, I concluded that the investigative purpose was to assign blame. The voluminous NTSB reports on the Aloha Airlines' crash did the exact same thing. Everybody simply came up with a list of people and other agencies to blame for what happened. The airport-security exercise was considered a success from the moment the young woman was targeted for blame and fired.

Once again, I thought about the list I had made in Nevada to evaluate and weigh potential causes of accidents. Human error had been at the top. At the time, I thought that human error caused systems to fail. I was wrong. Human error doesn't cause safety systems to fail; it's ineffective systems that cause humans to err. The suits at the airport were focused on the diagnosis, the agent, instead of the prognosis. To me, the prognosis was clear: to reduce human error, improve the safety system.

I failed to see how aviation safety could ever be improved to the degree it should be until the FAA focused on the prognosis (the system) instead of the diagnosis (assigning blame). I didn't yet realize that a climate of blame was as much a part of the structure of government agencies as two legs and two arms are to humans. I would remember all this many years later, when I challenged the FAA's dedication to safety and became a human sacrifice, just like the young woman in the airport.

Apart from my new insights, the airport exercise seemed to have little to do with training me to be an accident investigator and inspection agent for the FAA. It felt like a fraternity initiation, and, in a sense, it was. I was merely "the new guy" until that day. As soon as I returned to work the next morning, I knew that my status had changed. Properly trained or not, I now belonged.

"What's in the briefcase, Dave?"

"Hey, it's Dynamite Dave."

"I think you dropped your gun."

In addition to more jokes, a sign was taped above my office door: "New guys are a blast!" I took it down, but it didn't stop the teasing. That continued for weeks, until everyone got too busy with Christmas plans to care about teasing me. Everyone was busy but me.

Since I had neither the money nor the vacation time to see my family, I begrudged the holidays, feeling every inch like Scrooge. The night of our big office Christmas party, I felt so lonely that I even slipped out during the toasts to find a pay phone. Earlier, I had been coaxed into singing Christmas songs, and even though I had a chest cold that made my lungs hurt, performing had lifted my spirits momentarily, as well as further cemented my membership in the federal-employee club. But the good feeling had proven to be as fake as the snow around the Christmas tree in the banquet hall. I was deep into telling Jill that I didn't know if I could make it in Hawaii without them, when she said, "You don't have to." I thought she meant she wanted me to quit the FAA and come home. "Dave, I just heard from Starr Seigle McCombs. They want me to interview."

"Who?" She had said it as if I should know.

"One of the biggest advertising agencies in Honolulu. All you have to do is come get us, Dave."

Since I was moving my family to Hawaii, the FAA not only gave me time off, they paid for two airline tickets. We would still need a place to live and a car, so I got busy. I was living in a converted garage, and I now had seven days to fix up the makeshift apartment. No matter how hard I tried, it proved impossible. In the end, Tommy, my hippie landlord, graciously invited us to live in the guest bedroom on the main floor while I worked on the apartment.

Next, I found a deal on a 1975 Audi Fox. The eight hundred it cost ate up most of my money. The car ran great during the test drive, but by the time I arrived home, it was smoking like a mosquito fogger. I took it to a Samoan mechanic, who examined the engine. "Da water choked, and da head blown, bra." This translated to mean that the radiator had clogged and overheated the engine. "Somebody put sumptin in dakine." I didn't know what "dakine" meant, but I knew it wasn't good. The mechanic cleaned out the radiator. I spent my last two hundred, but I did drive the fogger home.

I flew to Denver two days before Christmas. I had a chest cold that was getting worse daily, but I loaded up on cold medicine and trudged ahead. We packed what we could and stored the rest. Since we had only two government tickets, I was flying back with Tyler, while Jill used a free-mileage ticket. She would arrive six hours later than we would.

By the time we boarded the flight, my chest cold was squeezing my lungs so hard that I wheezed and felt wobbly. I fell asleep as soon as I hit the seat, but I coughed so much during the flight that a flight attendant awakened me. She insisted that Tyler and I move to the mostly empty first class. The specks of blood on my shirt from coughing should have worried me, but instead, I spent the flight worried about violating FAA rules by accepting the upgrade.

When the bags and boxes arrived, I paid a skycap to stack them by the pickup lane. Instead of going to Tommy's house to unpack, as I had planned, I fell asleep on the bags to the sound of *Mario Bros.* on Tyler's Game Boy. Next thing I knew, Tyler awakened me by shouting, "It's Mommy! Daddy, wake up. It's Mommy!"

Oh shit! I had left a six-year-old alone in an airport for six hours. By now, I was sweating as if I were in a sauna and could barely move. I told Jill where the car was parked and gave her a map to the beach house. She would have to drive the mosquito fogger.

When we pulled into the driveway of the beach house, Tommy and his girlfriend came out to greet us. Jill took one look at Tommy's house, which was little more than a shack but still better than the garage apartment, and decided it would definitely not work for us. She explained to Tommy that I was ill, so we were going to stay at a hotel in Honolulu where I could be close to a doctor. Maybe they believed her, or maybe they knew it was an escape plan. Either way, I knew we wouldn't return to the beach.

I directed Jill to the Ala Moana Hotel, where some other FAA inspectors had worked out a reduced rate for employees. Once we entered the room, I tumbled into bed. I slept twenty-four hours straight before Jill awakened me and insisted that I go to the hotel clinic. The doctor examined me and quickly diagnosed pneumonia.

At first, I was even too sick to talk to Peter Undem on the phone, so Jill had to do it. Peter graciously advanced me sick leave since I had used up the few days I had accumulated. I would owe the FAA months and months of healthy service before I had any more time off.

After four weeks on antibiotics, the hotel doctor recommended that I get out of the room and into the sun to dry out my lungs. From that day on, the Hanauma Bay Nature Preserve became our daylight home. Jill and I listened to a local radio station on the boom box, while Tyler swam in the bay teeming with fish and turtles. He rarely left the water except to wander the beach trying to learn the local patois. After mastering a rhythmic pronunciation of the name of the state fish, humu-humu-nuku-nuku-apua'a, he earned the nickname "Fish Boy." Each day that we sat on the beach, looking out at Puowaina Crater, a site famous for human sacrifice, I had no idea that one of my coworkers was busy trying to add my name to the list of sacrificial victims.

CHAPTER FOURTEEN

Brothers in Arms

My first day back, I was catching up on a month of accumulated paperwork when Peter Undem summoned me. I figured he wanted to talk about the advanced sick leave and my obligations, but when he closed his office door, I realized it was something more serious. "We have a problem, Dave," he said in a dire tone.

"What is it, Peter, what's wrong?" My mind flashed on the worst-case scenario. Was I getting fired for missing work? Jill and Tyler had just moved to the island. We were broke. I was suddenly on the verge of a panic attack.

"Were you at the Haunama Bay beach a few days ago?"

"What? The beach?" I felt my fear turning to anger. "Yes, I was at the beach. On doctor's orders." Even as I said it, I realized how ridiculous it sounded—the doctor had prescribed a day at the beach! Then it hit me. How did he know I was there? "Who told you I was at the beach?"

"The doctor really ordered you to go?"

"He said it'd help clear my lungs to be out in fresh air, out of the hotel room," I said.

Peter nodded. "Then I'm sorry I jumped to conclusions, Dave. It's just that when Ron said . . ."

"Ron? Ron Willis*?"

"He was just trying to . . . he wanted to . . . you know."

"Yeah, I know. He wanted to discredit me."

"Well, being on the beach certainly gives the impression that you weren't very sick, doesn't it?"

"If you want to know why I was on the beach, Peter, just ask me. I'm happy to give you all the facts."

Peter's response was to feed pages from a file labeled "Soucie Sick Leave" into a shredder. It gave me time to think. Why would Ron Willis cause me problems? Both Ron and I, along with a few others, had been hired in the September budget rush. Some of the men hired, including Ron, seemed like poor choices to me. Were my feelings so obvious that he wanted to ruin me? Then I remembered hearing that Ron was upset when I got the Aloha Airlines' accident investigation. That must be it. To Ron, I was competition. Maybe I wasn't being fired after all, but I certainly was being schooled in bureaucracy. I was learning that the FAA is like the sea, where fish swallow other fish simply as a way of life.

When the machine finished shredding the files, Peter said, "Remember this, David. Even when you can prove you're innocent, the FAA can still take punitive action just for your giving the wrong impression. You better learn that and pay attention to impressions because they certainly do."

It was a rough way to return to work, but after I thought about Peter's advice, I decided to make an effort to counteract the damage that Ron had caused. I knew Ron was tight with Joe*, another recent hire. They shared some shady military connection. I could never remember Joe's last name and thought of him as Joe Friday. He was a dull and humorless sixty-three-year-old man who, because of his extensive military experience, could retire from the FAA after working only two years. Go figure!

Joe was a geographic-surveillance inspector. Not all the airlines flying in and out of Hawaii had their own island maintenance base, so the geographic-surveillance inspectors were responsible for observing and reporting on the quality of all contract maintenance work done on the island. Most of us in the Honolulu office were squeezed for space, but the five men assigned to geographic responsibilities had a large, open space filled with windows that overlooked the maintenance hangar and the airport tarmac and gates. It was down the hall a hundred feet from my own office, but I had been inside only once, during my orientation tour. On that visit, Joe and his supervisor had been busy investigating a small airline called Discovery. I didn't remember the details, but I clearly remembered Joe's words: "We'll get them shut down, all right. Those fucking gooks think they can come in here and run an American airline." I headed to Joe's office again, this time to check out my damaged reputation.

Despite energetic handshakes and a large dose of forced cordiality, I learned nothing about whether my reputation had been smeared or not from my second visit. Some men were friendly and some were jerks, same as before. Some asked me about my illness, and others didn't know I'd been gone. Nobody mentioned my recovery time on the beach. After a while, I gave up playing spy and wandered over to the windows overlooking the hangar. Below, the silhouettes of two beautiful, new planes and the name DISCOVERY reflected off the freshly painted white epoxy floor. Mechanics were removing an engine from one of the jets with surgical precision. A man stood watching the procedure with a stopwatch in his hand. "Why are they removing the engines, Joe?" I said. "Those jets are brand new."

Joe walked over to stand beside me. "Got nothing else to do," he said. "They do it every day."

After the engine was removed and put on a cart, the cart was carefully rolled to the engine-disassembly area, where it stopped precisely on four red marks on the floor. I opened a window in order to listen. "Seventy-three minutes, two-point-seven seconds," the man with the stopwatch yelled. "Our best time yet. So let's go again."

Moments later, I heard the hydraulic motors whine as they lifted the huge engines into exactly the right spot for reattachment. "Fuel, check. Electrical, check. Hydraulics, check. Mechanical, check. All connections go." It was like watching a crack drill team or marching band. It was truly impressive.

"Hey, Dan*, look! It's Mr. Soucie!" The work suddenly stopped while three of the mechanics looked up at me. My expression must have reflected my confusion. *Do I know them? Who are they?* "It's us, Mr. Soucie," the same man yelled. "From your systems class at Colorado Aero Tech."

I now remembered them from class. I smiled and waved and told them I'd be right down. When I got to the hangar floor, I recognized several of my students from Colorado Aero Tech. Dan, Jason*, and Rodney* were all in their early twenties. They told me they had been in Hawaii for about a year working for Discovery Airlines. They immediately launched into telling their fellow employees about my unorthodox teaching style. "Mr. Soucie would do whatever it took to teach us," Dan said. "There was this one time when I didn't understand how a VHF omnidirectional radio worked. I just couldn't get it, so he jumped on the table in front of me, spinning around like a top and beeping. I got it then."

I looked up and saw Joe Friday watching us. The words "giving the wrong impression" immediately came to mind. I couldn't appear to be too chummy. Neither did I want to be rude, so I arranged to meet the young mechanics that night.

We met at a bar far away from the FAA office. My three former students brought a fourth man with them. I remembered being introduced to him earlier in the hangar. Waylon* was ten years older than the Aero Tech graduates and had been at Discovery Airlines for much longer.

"Discovery is a great company, Mr. Soucie," Dan said.

"Dan, we're in a bar, not a classroom. Call me Dave. And yes, after watching you guys change that engine, I'd say you're right. Discovery must be a good company."

"Then why are you shutting us down?" Waylon had spit out his words with venom. The table got very quiet.

Dan quickly apologized for his coworker. "You don't talk to the FAA that way, man," he told Waylon.

The FAA? Oh yeah, that's me. "I heard you were being investigated, but I didn't know you'd been shut down."

"Yeah, right!" Waylon said with an Elvis snarl.

I tried to overlook his attitude. "For what reason?"

Dan started to explain, but Waylon cut him off. "You idiots at the FAA pulled our operating certificate. Said we're owned by a foreign national."

The United States forbids the foreign ownership of any domestic airline. I understood the reasoning. In time of war, the government can take over commercial airlines to use the equipment and facilities for defense purposes (the same goes for railroads and interstate highways). If the airline is foreign owned, this could pose a problem. But even knowing this, it still made no sense to me that the FAA would shut down an airline because of ownership issues after it was already operating. Ownership is confirmed before the operating certificate is issued.

Waylon was ready to explode, and I appeared to be the likely target. Dan stepped in. "Please understand, Dave, Waylon's losing his job. We all are." The happy Colorado Aero Tech reunion at the bar suddenly felt like a funeral gathering. "Think you could help us, Mr. Soucie?" Dan said. "Can you help save our jobs?"

The next day, I went to Peter Undem and asked about the status of Discovery Airlines. He explained that the Department of Transportation (DOT) originally had given Discovery approval to operate while the lawyers wrangled over the role of Philip Ho, but now the DOT had taken it away. "Philip Ho is Chinese American," Peter said. "I mean, he's an American citizen and also the primary stockholder."

"Then I don't understand the problem."

"Well, Ho's employed by Japan's Nansay Corporation, and Nansay's closely tied to the Japanese government."

"But if Philip Ho owns the airline . . ."

"Well, DOT lawyers figure that the Japanese government, not Ho, actually has operational control over the airline. And that's illegal. So they revoked the approval."

Peter took a file from his desk, while I thought about how to keep my promise to my former students. Despite my airport-security-test adventure, it was already clear to me that being an FAA inspector involved shuffling papers more than it did shuffling through molten aluminum at crash sites. Still, I remembered the satisfaction of cracking the Aloha Airlines' mystery. "Let me look into it, Peter," I said. "With my background in business—the aviation business—maybe I can come up with a solution."

"Sure, Dave, if that's what you want. Go ahead. I'll get you all the information. Only thing is, you're on accident duty today. I've got nobody else." He handed me the file.

I didn't even know what "accident duty" meant, but I was pretty sure I should be trained in order to do it. I figured Peter had forgotten that I had not gone to the training academy yet, but since I wanted to prove myself after the beach incident, I kept my mouth shut.

"A couple of small aircraft have failed to report," he said. He gave me the file. "Maybe drug planes or just a screwup. See if you can find them." I nodded and started to leave, but as I opened the door, Peter said, "And, Dave, this Discovery Airlines' business? Just remember where you are and who you're dealing with."

I was confused, and my expression communicated it.

"It's Hawaii, Dave. Pearl Harbor, remember? And Discovery is Japanese-owned." Peter shrugged.

To help my former students, I needed to discreetly assist Discovery Airlines, and to repair my credibility, I needed to find two missing planes. How hard could either task be? I nodded an okay to Peter and left.

CHAPTER FIFTEEN

Call Off the Search

Upon returning to my office, I opened the accident-duty file. The top page was a list of accidents, incidents, and anomalies picked up by air traffic control the night before. I searched for activities in the Hawaii area. Halfway down the page, I found what I was looking for: "An aircraft being transported to Tokyo, Japan from Oakland, California via Honolulu, Hawaii did not close flight plan."

As Peter said, it could mean anything. Sometimes planes secretly landed on remote dirt runways to load up with lucrative Hawaiian marijuana. Other planes simply changed their heading without telling anyone. Most likely, this plane never left Oakland. Pilots sometimes failed to cancel the flight plan when the flight was cancelled.

I scanned through several other occurrences and accidents before I reached the bottom, where I found a second notation: "An aircraft being ferried to Tokyo, Japan from Los Angeles, California via Honolulu, Hawaii did not close flight plan."

Two flight plans out of California left open on the same day? It could be a duplicate of the same incident, a typo, or a miscommunication, even though one said Oakland and the other Los Angeles. I had to find out if I was dealing with confusion or catastrophe.

My first call was to Tokyo, to the buyer of the airplanes. A Japanese man answered the phone. His English was broken, and my Japanese was limited to *sushi*. I explained why I was calling him, slowing down my speech, selecting easy words, and even speaking louder in that ridiculous way we all seem to do when confronted with language barriers. When I was done, he said, "They be here tomorrow."

"No, they won't be there tomorrow, sir," I told him. "They haven't even arrived in Honolulu yet."

"They be here tomorrow."

This continued until I finally gave up on Japan and phoned the airport in Oakland, the departure point, where I got in touch with a mechanic who had outfitted the plane with long-range fuel cells. "Oh, he took off from Oakland all right," he said. "Right on time."

Next, I spoke to a mechanic in Los Angeles. He confirmed that a second aircraft had left there at about the same time as its Oakland counterpart. The second aircraft had also been fitted with long-range fuel cells.

I made more calls and discovered that the pilots of both small planes had originally filed flight plans, but both had departed under visual flight rules (VFR). This happens when a pilot expects clear skies and doesn't intend to use navigational aids. The decision by both pilots struck me as a risky proposition for a transoceanic flight.

I got back in touch with the LA mechanic. "Why would your guy fly VFR? If something goes wrong, the pilot's got nothing but a wet compass to get him to Hawaii."

"Well, you know . . ."

"No, I don't know. So tell me. This is my job."

"GPS, man. He had a handheld GPS. The whole electrical system can fail, and he'll still get there."

The mechanic didn't want to admit that the pilot who left LA was using a Global Positioning System (GPS) unit for navigation. While standard today, in the early 1990s, GPS was not an approved navigation system, although pilots frequently carried a GPS unit as a backup for the less reliable Loran system then in use. The pilots argued that because the GPS unit was not installed in the cockpit, the unit was not a violation of FAA regulations. I didn't care one way or the other. "Well, GPS or not," I said, "the Los Angeles plane never arrived. Nor did the Oakland plane."

"Weird, huh?" he said insightfully.

Later that day, I gave my report to Peter Undem. "The two airplanes left California at about the same time, heading to Honolulu. Both filed flight plans but then went VFR, so they weren't on radar. I've checked out every arrival at every airport in the Hawaiian Islands. Nobody has them landing, and based on what I've learned, I don't think they were drug planes. There's no evidence of a screwup in record keeping. They just disappeared."

Peter stared out a window in silence. Finally, he said, "Dave, what you're telling me makes no sense. You're missing something. There's always an explanation."

I knew that Peter was right. All my life, I've seen how cause and effect work. There is always a reason for an aviation accident. "I'll keep looking," I told Peter.

I walked and hitchhiked back to our hotel that night, pondering what I might be missing. Jill had Pig Pen that day. Pig Pen had replaced Mosquito Fogger as the name for the old Audi. Smoke now encircled our car as it did the famous Charlie Brown character. Jill took one look at me and said, "Bad day?"

My first time on accident duty and I couldn't even complete the investigation form, because I couldn't find the missing planes to prove there were accidents. "Not if you enjoy being a fool. What about you?"

Jill was happy in her new advertising job. She smiled. "I found us an apartment today."

"Really? Now that is good news. Can we afford it?"

"Can we afford any place we'd want to live?" She had a point. "The realtor will meet us there so you can see it."

Half an hour later, we pulled up outside a beautiful high-rise condo at Banyan Tree Plaza. Thankfully, the real estate agent didn't see us arrive in Pig Pen, or he would have changed his mind about renting to us. "Nearly six hundred square feet of Hawaiian paradise," he said as he showed us around the tiny condo. We took it.

We were within walking distance to the beach, so the rent was twelve hundred dollars per month. Even with Jill's salary, we could barely afford it, and if we paid our bill at the hotel, we were short on the deposit. We wound up borrowing the money from my sister Grace and her husband, Kirk, even though they were pinching pennies as well. Their generosity saved my FAA career.

Their help also reminded me that I had promised to help my former students. So far, I had done nothing about looking into the problem facing Discovery Airlines, even though I had both training and experience in restructuring a company. While I was at Air Methods as the company expanded, I had helped our attorneys do this very thing.

The following morning, I contacted the Discovery Airlines' office and learned that Philip Ho, who lived in Asia, was currently in Honolulu. We met that afternoon.

Philip Ho was an older man with a sharp Asian accent. His small nose failed to hold up his glasses, and throughout the meeting, he kept pushing them up. Ho's American attorney, wearing a suit and tie, accompanied him to the Discovery Airlines' offices. It was clear to me that neither of them held out much hope for DOT approval. "Why not?" I asked.

The attorney responded, "Because the DOT thinks Mr. Ho is merely a figurehead, and that Nansay will be in control."

"Is that true?"

Mr. Ho didn't even bother to look at his attorney. "In the Japanese culture, you do not dishonor your boss."

I appreciated his honesty, even though it meant that Discovery Airlines likely was doomed. Despite this, I spent hours with the two men, going over possibilities. We came up with three different scenarios to present to the DOT to get the agency to overturn its decision.

Afterward, when I ran into any of my former students, they would ask me if I had heard any news. Before long, I got tired of responding, "Not yet," and tried to avoid them, although, from time to time, I still watched the mechanics practice removing and installing jet engines.

I was admiring another precision engine-replacement drill the day Joe Friday told me the news. "The gooks lost," he said. He looked at me as if trying to get a reaction. All I did was nod and walk out the door. The next day, Discovery Airlines, with the finest maintenance organization I would ever encounter, went out of business.

I didn't hear from any of the young mechanics for weeks. I figured they either were disappointed in me for failing or didn't believe I had made a sincere effort to help. But one night, all three of them showed up at our apartment with a huge bag of fresh lobsters. "At first, we were really bummed," Dan told me, "but then we got hired as deckhands. And now we live near the beach and eat lobster." After that, the lobster fest became a regular event when the crew returned from a fishing trip.

My life with the FAA became routine again. I knew it would remain that way until I got to Oklahoma City to complete my training to become a full field agent. Meanwhile, I tried to ignore the jokes at my expense. My failure to save Discovery Airlines was one thing, but my inability to find two missing airplanes was the stuff of legend, and not one I wanted. The two planes had disappeared on April 5, 1991. A month later, I still had no leads.

Tyler met me at the door one May evening when I came home. "Let's go swimming, Dad." He looked up at me through a snorkel mask he had put on crooked. No matter how difficult my day, Tyler always put a smile on my face.

I put on my swim trunks and snorkel gear. As my son and I flapped our way out the door, Jill yelled, "Don't forget, the space shuttle launch will be on TV again tonight." Tyler and I loved to watch space launches.

After our swim and dinner, we sat in front of the TV, listening to Tom Brokaw's penetrating voice and watching as the cameras showed steam around the shuttle and ice falling from its sides. "This mission has certainly had its share of setbacks," he said, "but now we're at T minus two minutes, and everything is go for the space shuttle *Discovery*."

Discovery? The name of the shuttle had to be *Discovery*? Life was tormenting me.

"This mission was originally scheduled to launch before the *Atlantis*, which was launched on April fifth."

I remembered watching the *Atlantis* launch, but not much else about the mission. Tom Brokaw came to my rescue. He explained that the *Atlantis* mission had been used to calibrate the navigation instruments located at the Air Force Maui Optical Station on the Haleakala Crater in Hawaii. Wow! Right next door!

I watched the replay of the morning liftoff with as much attention as I could muster with my young son in full Batman regalia bouncing around at my feet. After it was over, Jill said, "Did you figure it out yet?"

"Figure what out?"

"Whatever you've been thinking about all evening."

I grinned at her and said I was going for a walk. It was what I often did when I needed to solve a problem.

By the time I reached the beach, the sun was settling into the ocean and the gloaming had begun. While I walked, I kept replaying Tom Brokaw's words in my mind. Why? Why were they stuck in my head like a summer song? Why were the words *"Atlantis" and "Hawaii" and "Haleakala Crater" and "calibrating navigational instruments"* so important?

"A month earlier," I suddenly said aloud. What had happened a month earlier on the fifth of April? The fifth of April was the day of the *Atlantis* shuttle launch; the fifth of April was the day NASA calibrated the navigation systems for the Air Force; the fifth of April was the day the two planes using GPS left California and disappeared. Could there possibly be a connection?

I ran back to the apartment and burst through the door. "I'm going back to work," I said breathlessly. "Something I've got to check out. I may be all night."

"So you figured it out?" Jill said.

"Oh yeah, I figured it out."

"Do you want my cape, Daddy?"

"No thanks, Tyler. I think I can do this on my own."

"I'm not Tyler," my son said. "I'm Batman."

Two days later, I hurried into Peter Undem's office. "I found 'em," I announced. "Here's the accident report."

"Found whom?" I knew that Peter had written off the missing planes as a couple of drug runners avoiding detection. In fact, he had reached this conclusion before even assigning me the task of locating the planes.

"The two planes that left California and disappeared."

"Really?"

"The planes left on April fifth, the same day NASA shut down access to the GPS satellites while the *Atlantis* shuttle was over Hawaii.

Both of the pilots were using handheld GPS units for navigation. When the GPS stopped working, the pilots had no idea where they were. I think they just flew around over the water in circles until they ran out of fuel."

"And crashed in the ocean?"

"And crashed in the ocean. The impact probably killed both pilots. Or they drowned when the planes went under."

"That would definitely suck."

"Two small planes lost somewhere in a big, big ocean."

"They could easily disappear." Peter looked up over his desk lamp. "Makes sense. Just put it in my basket."

"Put it in your basket? Seriously?" I couldn't restrain myself. "I put a lot of work into this, Peter."

He got up and walked over and patted me on the back. "Is that better?" When I saw the grin on his face, we both chuckled at how I had reacted. "Sit down, Dave. I have something to tell you." He closed the door.

Uh-oh! The last time he did that, it was bad news. "What'd I do now?"

"An excellent job, actually," Peter said. "And that's why you're going to Oklahoma. Two slots just opened up at the training academy. You leave the end of July."

"Fantastic!"

"I thought you'd like that part of it."

"What isn't to like?"

"Well, the other slot is going to Ron Willis."

CHAPTER SIXTEEN

Rock Fever

A few days later, I got the travel orders for Oklahoma City, but the elation I felt in Peter's office had already been replaced by anxiety. I had accepted the FAA job in Hawaii without truly consulting Jill, and then I had encouraged her to uproot Tyler and herself to join me, and now, after six months together on "the rock," I was going to leave them again. On top of that, Jill had been unusually cranky for a couple of weeks. I avoided telling her about Oklahoma City for as long as possible.

Sam Matsumoto, a pilot and senior FAA inspector, asked me to leave work early with him to help install a new garbage disposal at his home. By the time we reached his condo, the island had lost power. It happened from time to time. Since our own rental condo was on the thirty-second floor, it would be a long climb without an elevator. I called Jill to see if she and Tyler were all right, but she didn't answer.

"So how does Jill like it here?" Mrs. Matsumoto asked me once I was off the phone. She was setting tofu and something resembling chopped rabbit food on the table.

"Well, she loves her job," I said. "But living here, I guess I don't know. I've never really asked her."

Mrs. Matsumoto politely didn't pursue the issue, and I was left, instead, to contemplate my own selfishness. The power returned long enough for me to replace the disposal, but by the time I got home, we were in the dark again.

I walked by the pool on my way inside and found Jill sitting there in a nice work dress with her shoes off and her legs dangling in the water. She was watching Tyler swim. "Where've you been?" she asked sharply.

"At Sam's, fixing his garbage disposal, remember?"

"No, I don't remember, David."

"What's wrong? Why are you wearing your work clothes?"

"What's wrong is that I had to walk down thirty floors from my office downtown wearing sandals just to get home." Jill flung the offending sandals into the swimming pool. "But I guess I should've walked up and down another thirty-two floors to change, so I'd look pretty out here by the pool, right?"

Jill never talked like this. "Are you okay?"

"What do you think?" She got up and stomped toward the building to make the long climb up to our rental condo.

The next morning at work, I was still pondering Jill's personality change, when I heard, "What are you doing here, Dave?" Peter Undem was standing in my doorway.

"Well, where should I be?"

"In Maui with Wayne." Wayne McKinney was an inspector hired a year before me. "A parachute jumper died over in Maui," Peter explained. "I want you to go with Wayne. Take the G-car and catch him at the airport."

He threw me the car keys, and I raced to the airport and managed to catch up with Wayne in the parking lot. He drove a small convertible, and he was so tall that his head stuck high above it. That's how I spotted him.

Our puddle jump to Maui found us at the Wailua Airport before noon. Three hours earlier, the plane had been in an accident that resulted in the death of Bill Boyd, a marine. Wayne immediately separated the witnesses from everyone else, then questioned the policemen and firemen at the scene. "Where's the body?" he said.

"Coroner has it."

"And Sergeant Boyd's chute?"

"In the red truck over there."

"I need a certified parachute rigger to look at it," Wayne told a policeman. "Find me one." Wayne never raised his voice or changed his calm tone, yet he took charge. I was so in awe of his skill that it took me a while to realize I was working my first official accident site.

Moments later, we walked over to the pilot. He was holding the right door of the plane open, while his foot was perched on the landing gear in a confident pose. Above him was a hole, two feet wide and two feet deep, near the front of the wing. An unopened parachute lay on the ground. "You Wes Crey*?" Wayne asked.

The pilot answered with a grin and a nod.

"I'm Wayne McKinney, the FAA inspector in charge, and this is Inspector Soucie."

"Sushi?" the pilot said, grinning even wider.

"It's Su-Sea, Inspector Soucie," Wayne said gruffly; the pilot's confident attitude suddenly changed. Wayne opened his dark green government-issue notepad, clicked his blue FAA pen a few times, and said, "What happened?"

"Happened? Well, I took up these four guys who wanted to jump during the eclipse, you know." *Click, click, click*. "The eclipse of the sun this morning. You saw it, right?" Wayne didn't even look up. *Click, click*. "We loaded up and . . . and they all climbed out on the wing, and then some shit went really wrong, and I, I . . ."

Wayne clicked his pen rapidly and turned to a policeman standing nearby. "All the jumpers here?"

"Not the one who . . ."

"Other than the deceased?"

"Uh, the seaman, he's not here," Crey blurted out. "He had to report back to his sub there at Pearl."

"Let me talk to the other two," Wayne told the cop. "I'll get back to you, Wes." As the policeman took the pilot away, Wayne whispered, "The seaman's all yours, Dave. I won't go on a submarine. I'm claustrophobic."

Wayne interviewed the two jumpers and a few people on the ground who had seen Sergeant Boyd's fall. During the entire process, he said very little. He mostly clicked his pen and jotted notes in his book. When he was finished, I returned to the plane and examined the hole in the wing. It was hard to believe that an airplane could have landed with such a gaping hole. It reminded me of my ride with Little John and his expert landing of our damaged airplane.

"Okay, Dave, here's what we got," Wayne said after joining me. "The plan was for the four guys to jump in tandem during the solar eclipse, three of them joining hands in a circle with the solar eclipse behind them while the other took photos. Okay? So the deceased climbed out on the wing. He's got the camera. Now two of the jumpers followed him out, but the seaman was behind them in the plane. Then something caused Boyd's drogue chute to open, and it caught on the front of the wing and took him with it. This caused his main chute to open, and it"—Wayne pointed to the hole—"it pulled Boyd through the wing right here." He paused and picked at a loose piece of sharp aluminum. "Killed him instantly and pulled the wing back, causing the plane to flip upside down. Two jumpers then bailed out and landed safely."

"Two? So the submarine guy stayed in the plane?"

"For a moment. Then he jumped. Sound about right?"

I looked at the damaged wing. "Yeah. It sure does."

"Good. Now, let's see if we can figure out why Boyd's drogue chute opened. Crawl up there on the strut."

The strut started wide and narrowed as it went up to the attach point on the wing. I perched on it, facing backward, then reached up and grabbed the front of the wing to pull myself out farther until I was under the hole.

"God, Dave, you're bleeding." Wayne was looking at my hand, covered with blood and gore. It was odd. I hadn't felt anything, not a jab or a cut. "Get down from there."

I tried to slide off the strut, but my shirt snagged on something. The wing above me had an air scoop still attached to a metal vent. It had hooked my shirt collar right where Boyd had been pulled through the wing.

After Wayne freed my shirt, I jumped down, wiped my hands, and looked for a cut. There was none. My hand had been covered with the blood and body tissue of the dead jumper. I wanted to run somewhere and sterilize my hands, if not my whole body. Instead, I picked up the parachute that the pilot had left behind and held it up next to the airplane vent. There are three chutes: the drogue, the main, and the reserve. The lanyard to deploy the drogue chute was perfectly in line with the vent. When I looked inside the wing, I found a frayed and bloody piece of a release lanyard. I showed Wayne.

"Damn!" Wayne said. "I think you solved the mystery."

That evening, instead of telling Jill about Oklahoma City, I rattled on about my first experience investigating an accident. I didn't go into all the gory details, but I hinted at them enough to hold Tyler captive. Throughout it all, Jill seemed distracted.

When I was ready to leave for work the next morning, she told me she was going to call in sick. I was torn between staying home with her and visiting the submarine to interview the remaining parachute jumper. Wayne and I had a theory about what had happened, but I needed to verify it. Since it was the submariner's last day in port, I decided to go early and try to hurry back home.

When I arrived at the Pearl Harbor naval base, the submarine commander met me on the dock and explained that the seaman I wanted to talk to had to remain on ship. "You can use my quarters for the interrogation," he said. I climbed down the ladder and into another world.

"Welcome aboard, sir," the ship's steward said. After an awkward moment of our trying a half salute before shaking hands, he explained, "This is the USS *Houston*, a Los Angeles class submarine, powered by an SG6 nuclear reactor. We house as many as twelve Tomahawk T-L-A-M missiles." Nobody had said anything about missiles!

The commander interrupted, perhaps feeling that I was getting more information than needed, and told the steward where to take me. We walked through a maze of tight halls before I was left alone in the commander's quarters.

The skydiving seaman arrived quickly and introduced himself as Taylor Hendricks*. I noticed that he sported a black eye. Imitating Wayne, I pulled out my dark green, government-issue notebook and blue FAA pen. *Click, click, click.* "Now tell me what happened."

"The guy fucking hit me in the eye, that's what."

"Hit you in the eye? Boyd?"

"The pilot. After he landed, he sucker punched me."

Curiosity made me want to pursue this story, but I knew I had to have Hendricks verify everything. "Start from the beginning," I said.

Hendricks told me that the solar eclipse that spurred the skydiving adventure was so rare that it happened only once every 360 years and lasted for no more than seven seconds. The skydivers had carefully orchestrated the photos they planned to take. "I love to skydive, but this was something special. I knew I was lucky to be in Hawaii at this time." *Luckier than Boyd*, I thought.

He went on to explain what had happened in the air. Mostly, I just listened and clicked my pen and took notes, sometimes asking questions for clarification. Many times, I found my mind drifting from his story to Jill, wondering what she was doing at home. When Hendricks got to the part where his jump buddies bailed out, he had my attention. "Because of the damage to the wing, we were flying upside down, so I told the pilot to jump," he said. "He had a chute on and everything, but he didn't do it."

"And you did?"

"Of course. I skydive because I like some danger in my life. You get bored on a sub. All this structure. I mean, I like a little danger, but I'm not stupid, so I jumped. No way the pilot could fly that plane."

All in all, Hendricks's story confirmed our theory: Marine Sergeant Bill Boyd's ripcord had gotten caught in the scupper when he slid out onto the wing, leading to a series of rapid events that killed him and crippled the Cessna 185. Of course, Hendricks was wrong about one thing. The pilot was fearful of abandoning the crippled airplane and having it crash in a populated area, so he stayed in the cockpit and miraculously managed to land it.

"One more thing," I said. "How come Crey hit you?"

"'Cause he's a jerk." *Click, click.* The seaman shifted his weight and clenched his jaw. Finally, he said, "Because we all jumped out and left him to die."

Wes Crey had come off as smug and unlikable, but his heroic actions made me change my mind about him. I'd make sure his actions were noted in my report. We were done, so I asked Hendricks to escort me off the sub. I found it strange that he didn't find sleeping beside nuclear bombs dangerous, having to go skydiving to find excitement, but I sure couldn't wait to hear the sound of shoe leather on land. On this visit, I began to understand that recognizing risk was a matter of perspective. Like the frog that boils to death in slowly increasing water temperature, Hendricks had no concept of the risk of being on the nuclear submarine.

I didn't make it back home, but later that day, Jill phoned me at work. "I went to the doctor," she said. "He told me I have 'rock fever.' He said it happens to some people here—this feeling like you're stranded with no way out." I started listing the virtues of life in Hawaii, but Jill cut me off. "Dave, there's something more we gotta talk about, but we'll do it tonight, in person."

"Okay," I said and then spent the day wondering if my marriage was in trouble or if Jill was seriously ill or . . .

That night, after leaving Tyler with a babysitter, we walked to a restaurant where Jill had made a reservation for a secluded booth. Whatever she wanted to talk about, I knew it was important. We had barely sat down and ordered drinks before Jill blurted out her news. "I'm pregnant!"

CHAPTER SEVENTEEN

Paradise Lost

"I've suspected it for a while," Jill said, "but I wasn't even sure I was going to tell you."

"How could you not tell me?" I said. "Don't you think I would've noticed?"

Jill began crying softly. "I just didn't want to tell you. I want the baby, but we can't afford to have it here, Dave. Not if I can't work." I wanted to tell her she was wrong, that everything would work out, but all I could do was hold her. "When I tell the ad agency, they'll fire me. They as much as told me that when I got hired." Her eyes were cast down and her chin trembling. "I'm sorry for keeping a secret from you."

I held her chin and lifted her face. "You don't have anything to be sorry for. Besides, I have my own secret."

"What do you mean?"

I finally told her about going to the FAA Academy.

"Shit, shit, shit." More tears flowed down her face. "Things just go from bad to worse, don't they?"

"Just tell me what you want, Jill."

"I wanna go home," she said. "I'm so scared being here on the island. I feel like I can barely breathe." Jill didn't like island blackouts and being separated from her family. In short, she hated living on an island.

The next day, I scoured the personnel manual, looking for a way to get reassigned to Denver. Any hopes I had were dashed as soon as I read the section on training: "After completion of training at the Academy, the employee must remain at his current assignment for no less than one (1) year." If I went to the academy now, I would be ineligible for a transfer, and if I refused to go, I still wouldn't get a transfer. There was only one solution.

"I'm going to quit the FAA," I told Jill that evening.

She listened to all my reasons. When I finished, she reminded me of why I was with the FAA—I wanted to make flying safer. I believed that I could. When I insisted that quitting was the only answer to our problems, she encouraged me to at least complete the training. "It'll look good on your résumé," she pointed out.

A few days later, Tyler and I accompanied Jill to the doctor's office for her first pregnancy checkup. It was a beautiful and exciting day. The nurse greased up Jill's belly and rotated the ultrasound wand across it.

"Is that it?" Tyler's eyes were glued to the display.

The nurse didn't reply. Instead, she set the ultrasound wand on the table, asked us to wait, and said she'd be right back. Jill looked up at me. "This isn't good."

I squeezed her hand and held it while we waited. When the nurse returned, Jill's doctor was with her. He was young and handsome and spoke softly. "Nurse, would you show our dashing young man here to the playroom?" Tyler reluctantly went with the nurse, while the doctor moved the ultrasound wand over Jill's abdomen again. I could see her concern intensifying. After a few minutes, the doctor wiped the gel off her belly and took her hand from me. "I'm afraid there's no heartbeat, dear."

Here it was again, this sense of cosmic injustice that causes humans so much pain. People who don't want kids too often have them, while those who want them too often suffer disappointment. For us, it was the second child we had lost this way.

Over the next few days, our condo was unusually quiet as Jill's irritability turned into depression. I couldn't take it. I was willing to give up a lot for a career in the FAA, but not my wife. "I think you and Tyler should come with me to Oklahoma," I said to Jill. She started to protest, but I wasn't done. "And after that, you'll go on to Colorado. I'll move back there as soon as I can."

The effect was like opening the curtains to let the sun brighten a dark room. A flurry of action followed: I sold Pig Pen, we broke our lease and forfeited the deposit, and Jill told her employer the news. In the end, all that mattered was that we were leaving Hawaii together.

Driving into the Mike Monroney Aeronautical Center (MMAC) in Oklahoma City the first time was overwhelming. Instead of the "little red schoolhouse" I expected, the aeronautical center is somewhere between China's walled Forbidden City and a military base. It's both massive and bustling.

Adjacent to the Will Rogers World Airport, the MMAC actually consists of five different departments: the Civil Aerospace Medical Institute, the Logistics Center, the Civil Aviation Registry, the

Mike Monroney Aeronautical Center

Transportation Safety Institute, and the FAA Academy, where inspectors, air traffic staff, flight attendants, pilots, mechanics, and almost anyone else working in aviation go for training.

I started my first class a couple of days after we arrived. Jill and Tyler settled us into a furnished rental apartment in a compound used almost exclusively by FAA trainees and their families. Jill's mood had lightened considerably since leaving Hawaii, but I would still catch her staring off into space at times, looking sad.

Early on, the academy administration decided that I would not follow the usual five-week training. Because I had extensive aviation experience, I was slotted into a different program. I would attend school for four weeks, have a two-week break, and then return for three more weeks of training. After graduation, instead of simply being certified in general aviation—small planes, helicopters, charters, mechanic schools, etc.—I would also be qualified to do oversight on major commercial air carriers.

Many of the classes were tedious and taught by instructors who relied heavily on slide shows. In my regulations class, we went through the entire CFR 14, Code of Federal Regulations. The class in certification taught us how to evaluate applications for an operating certificate. I wished that I had taken this one before trying to save Discovery Airlines. I attended classes on ballooning, amateur-built aircraft, and a variety of other topics. The two courses that I most looked forward to were on surveillance, in which we learned how to conduct ourselves in the field and how to plan surveillance, and on accident investigation, in which we studied pieces of broken airplanes and endless hours of slides that showed mangled bodies and demolished aircraft. Since I had already experienced accident investigation firsthand, the class proved disappointing. Over the years, I returned to Oklahoma City many times for advanced courses in accident investigation involving both small and large airlines. These classes rarely disappointed me.

Most academy classes lasted three to five days. Students had to pass a final exam in each class before advancing to the next class. Failing an exam meant you were washed out of the FAA, so the pressure was intense.

During an extended holiday break from the academy, I took Jill and Tyler home to Colorado and then returned alone to Oklahoma for the final three weeks of training. I had no idea how long it would take me to find a new job in Colorado or, if possible, to find a way to get the FAA to transfer me to the mainland. Jill and I decided I would go back to Hawaii alone for as long as I could stand to be apart from her and Tyler. I had no idea what was ahead.

When I returned to Honolulu, Peter informed me that I still had to complete a list of specific duties as apprentice before I could become a journeyman inspector. A senior inspector had to observe me performing all the required tasks and had to sign off on them, so I was assigned to shadow Steve Dahlen. True to the nature of the FAA job, most of the duties were clerical. Finally, after weeks, I had successfully completed all the required tasks except for one: "perform an on-site aircraft accident investigation."

My work on the parachute-accident investigation did not qualify, since, at the time, I had not yet attended the training academy. By October, I was desperate to complete the last requirement, so I examined the daily accident reports like a vulture looking for a roadkill. Peter Undem and Pete Beckner were just as anxious for me to become fully qualified. They were shorthanded and under a press and congressional microscope ever since the FAA shared the blame for the October 1989 crash of a commuter plane on Molokai that had killed twenty, eight of them members of a high school volleyball team.

October is a busy time for helicopter tours of the Hawaiian Islands, so it wasn't too surprising to eventually find a helicopter crash in the daily report. "I found one, Steve," I said, with more glee than I should have. "A tour helicopter got too close to the active volcano and crashed into it on the Big Island."

Within seconds of my announcement, Steve was standing at the door with his field bag in hand. "What are you waiting for, Soucie? Let's go."

A short time later, we arrived at the Hilo airport, where we boarded a helicopter and took off for the short flight to the accident site. As we descended to land, I could see the crash site was very close to the mouth of the volcano. The tail of the crashed helicopter had been severed, and the fuselage lay on its side, battered and beaten. Miraculously, there were no fatalities. All six passengers and the pilot had been airlifted to safety.

When our helicopter touched down at the site, steam and smoke billowed up and surrounded the helicopter. "Get out fast, the lava is unstable!" the pilot told us. We grabbed our gear and jumped out. As the helicopter lifted off without us, we could hear the lava pad under our feet groan and then lift about six inches.

One of the things emphasized over and over in the accident-investigation class was to get a good overview picture of the accident site from the approach path to the impact. To do this, I climbed the rocks along the approach path toward the highest point. As I climbed, I stopped several times to look back at the accident site through the camera viewfinder. I continued climbing higher toward the mouth of the volcano to be sure I got the best shot of the site. At the top, I checked my viewfinder again. One step back and I'd have a great shot. I had just lifted my foot when the hairs on the back of my neck stood up. I looked behind me and saw that I was literally teetering on the edge of the volcano. When I put my foot down to gain my balance, I dislodged a small rock and watched it fall some thirty stories into the roaring lava. A red mist of poisonous gases and steam rose up out of the volcano and came toward me. It seemed as if I had angered the goddess Pele.

I dropped my camera, abandoned my radio, and ran down Mount Kilauea toward the site as fast as I could, jumping from rock to rock with the red mist and hot lava at my heels. When I got to the site, I quickly grabbed the radio from Steve and called the helicopter for an emergency evacuation. The helicopter pilot had seen that the wind was now blowing the hot volcanic gases toward us and was already on the way when I radioed for help. When he touched down, we immediately threw our field bags and our bodies onto the helicopter floor. Before we could even get seated, the pilot lifted us to safety. Below us, I could see the crashed helicopter and the site where I had sacrificed my camera and radio to Pele disappear into the ominous red mist.

Steve, always the consummate professional, had taken the pictures we needed for the investigation, so my flight of panic hadn't caused us any real problems. Steve signed off on my last on-the-job-training requirement as soon as we got back to the office.

I was now a journeyman inspector with the FAA.

CHAPTER EIGHTEEN

See Rabbit Run

Peter Undem loaded me up with work as soon as I was certified. My inspection list included fourteen maintenance and repair stations, a mechanics' school, five charter-airplane companies, and two helicopter operators.

One of the helicopter operators scheduled for a routine inspection in a few weeks was Island Heli*. Tommy Cho*, the company owner, had a reputation for being tough. "Island Heli's a licensed tour operator," Peter explained, "but its real business is search and rescue, firefighting, and government work." Peter didn't explain what government work the company did, and I didn't bother to ask. "They have a great safety record," Peter added, "but we have a complaint to check out."

"Okay. What am I looking for?"

"An ex-employee says Cho is flying one of his helicopters without replacing time-limited parts. The caller didn't leave his name, so it's probably nothing, but we can't risk ignoring it."

I knew the risk Peter was referring to was due to the current scrutiny of the Molokai accident that had killed schoolkids. Since the crash, anything that could land the FAA in the news, even a complaint, had to be dealt with immediately.

It was still morning when I drove up to an eight-foot-high chain-link fence that surrounded Island Heli. A man leaving the hangar area was locking the gate behind him. "Hey, leave that open, will you?" I called out. The man ignored me and continued to wrap the chain around the gatepost. "Hey, I said leave it open." I got out of my car and walked toward him. "My name's David Soucie," I said, "and I'm with the FAA." When the man heard this, he dropped the chain and lock and ran to his truck. After looking back over his shoulder at the hangar, he hit the gas and raced off. I choked on a cloud of dust, while the gravel his tires threw up pinged against my shiny, new, black G-car. "Hey, you friggin' jerk! Get back here!"

I floored the gas pedal and sped through the gate. The man's actions had left me angry and out for blood. That was not good. I was supposed to arrive impartial and with an open mind. I repeated these words like a mantra as I parked in front of the largest of three hangars. A few other cars were parked there. All the hangar doors were closed, which was odd. In Hawaii, you never see the hangar doors closed. The weather is simply too nice.

I pounded on the office door several times, but got no response. I could hear movement inside, so I knew someone was in there. I walked over to the hangar doors, looked through a crack, and saw a man walking by. "Hey you!" I yelled.

"What?"

"I'm with the FAA and I'm here to . . ."

"Help me?" The man laughed. "What a crock of shit."

"Well, that may be true, but I have to examine the records of one of your helicopters."

"You can't," he said. "This is private property."

He was right about one thing—it was private property. But he also was wrong, very wrong. An FAA inspector has the right to enter any property at any time if the company charters to the public and repairs its aircraft there. I didn't even need probable cause. I explained this to him through the crack in the door. At the time, no other federal agents had the right to enter private property without first establishing probable cause. This was, of course, before the Patriot Act, which basically gives federal agents free rein to do anything if they suspect foul play. In 1991,

FAA inspectors actually had power that made the DEA and the FBI drool. "You can either let me in right now," I said, "or I can request the local authorities to join us. Your choice."

"Okay, okay, enough already. Spare me the lecture. I'll let you in." The man mumbled something about entering at my own risk as he unlocked the door. "Happy now?"

I thought, *Do these guys take a class in how to piss off the feds?* "I need to see your boss," I said sharply.

The man led me to the office, where I found a gruff-looking man in a black leather jacket, his hands clasped behind his head and extra-large boots splayed across a desk. "Well, hello there, Inspector Soucie." I couldn't imagine how he knew my name. I was sure that I'd never met him. "Have a seat," he said. "What can I do for you?"

"I'm here to see Tommy Cho."

"You're looking at him."

I sat in the chair closest to the door, put down my briefcase, and pulled out my green notepad and blue FAA pen. It was obvious that the clicking-pen method wasn't going to intimidate this guy, so I got right to the point. "You've been accused of flying helicopters that aren't airworthy. According to an accusation, time-restricted components aren't being replaced."

He snickered. "Accusation?"

I launched into the specifics of which helicopter and components I wanted to check. I was just getting warmed up when Cho stopped me. "Right here on my desk is the file. Copies of all my component times and receipts."

This was bizarre. He already knew my name and why I was there. Still, I picked up the file and began reading. Cho sat there and watched me, while the man who had brought me to the office drifted away. After a few minutes, I said, "Well, I need to ask you about—"

"Let me show you something," Cho said, "before we talk." His feet came off the desk, and his body sprang up. He opened a desk drawer and rummaged around in it. I returned to studying the complex helicopter-component time chart.

Cathunk! The sound made me look up. An Uzi, or what I took to be an Uzi, was slammed down on the desk. He spun it around until the barrel was pointed directly at me, and then he leaned across the desk. "Now we can talk."

I backpedaled in my chair until I hit the wall. I was holding the file folder over my chest as if it were a magic cape capable of stopping hollow-point bullets. "What the fuck is wrong with you, man?" I shouted. "I'm a federal agent. You can't pull a gun on a federal agent."

No sooner were the words out of my mouth than I heard helicopters approaching in the background. I figured they were some of Cho's fleet. *Just what he needs*, I thought, *reinforcements*. Cho heard them too. He jumped up and looked outside. "Shit!" he said.

With him momentarily distracted, I burst through the open door and ran to my car. I didn't waste any time in turning it around and speeding away. Fortunately, the gate was still open. About the time I gunned my G-car through it, a line of black SUVs roared toward me, while overhead, a swarm of helicopters descended on the hangar.

"What the hell?" Was this some movie set I accidentally wandered onto? If so, the SUVs looked real enough to be official government vehicles, the kind favored by the DEA and the FBI. I looked into the sky. The camouflage helicopters were unmarked, so they clearly were not Cho's. Any minute now, I expected one of the black SUVs to pull sideways and block my exit, but it didn't. In my rearview mirror, I saw men bolt from the cars with drawn guns. "What the . . . ?" I had to get out of there. I jammed the gas pedal to the floor and raced away.

I was shaking all the way back to the office. When I arrived, I rushed into Peter Undem's office and told him what had happened. It was clear from his expression, somewhere between dazed and dazzled, that he had no prior awareness of any raid planned on Island Heli. "You're telling me they arrived right after Cho pulled a gun on you?"

"An Uzi, I think. I've seen them in movies."

"Well, it's either one hell of a coincidence in timing or, whoever these guys are, they had his office bugged."

"You think it was a drug bust? Is he a gunrunner?"

"Had to be DEA and FBI. Maybe even spooks."

"Spooks? You mean like CIA?"

"Could be CIA, NSA, military intelligence. Whatever it was, this is big, Dave. This is big."

I showed him the file I had taken with me when I fled. "I looked at all his receipts and records on my return flight. Well, at least when I could stop shaking, I looked at them." Joking made it feel less real. "I found a receipt for a time-limited, flight-control lever. It should've been replaced at twenty-five thousand flight hours, but it was backordered. So we can prove he knew the part was overdue for replacement and kept flying the helicopter anyway." We both understood the meaning. With this evidence, Cho not only was looking at a fine and suspension of his flight authority, but he was also facing possible criminal charges and imprisonment.

"Gees, Dave, this really is big." Peter patted me on the back. "It'll mean a court hearing. You gotta really prepare. Cross your t's and dot your i's, all of them."

Preparing an enforcement file for court takes a lot of time; I had plenty of that. Neither my request for a transfer to Colorado nor my job search in the private sector was going anywhere. So far, I had been able to fly back to Colorado often enough—doing "en route inspections" of commercial airplanes—that Jill and I were surviving. Otherwise, my only escape from loneliness was work.

Once the enforcement report was finished, I submitted it to the FAA Regional Headquarters in Los Angeles. Two weeks later, a young attorney from FAA Legal Enforcement paid Peter and me a visit. She introduced herself as Paula Merker*. Paula, a poster girl Jewish American princess, was wearing a business suit more common to Beverly Hills than Honolulu. "I'm here to prepare you for the hearing," she said.

Paula outlined the case and assured us that the regional office would recommend fines and a suspension of Cho's operating certificate. She didn't say anything about criminal charges, and nobody mentioned the assault on Island Heli that I had witnessed. When she finished her review, Paula said, "David, I need a word with you alone." Peter glanced at me, shrugged, and left the room.

"What's up?" I said.

Instead of responding, Paula got up and closed the door. By now, I was like Pavlov's dog—a closed door was a signal for bad news. "I want you to know, you did a good job on the investigation and report. But there's another matter we need to discuss." She stopped and peered out the window. "You know what? Let's get a cup of coffee or something. I think I saw a diner just down the street."

My mind was busy searching for what I had done wrong, but I couldn't think of anything serious enough to merit a closed-door session. I decided to remain silent; no use running toward the noose, I thought. This resolve lasted until we were seated in the diner. "Well, what is it?" I said.

"Our office has been made aware of a potential danger to you and your family."

I laughed. I couldn't help it. "You're kidding!"

She didn't laugh. Instead, her eyes scanned the room. "I need you to meet someone," she said quietly.

"Who? Why?"

"Someone who can protect you."

"Protect me from what? Who? Why?"

I had raised my voice, and she motioned with her hand for me to be more quiet and did another sweep of the people in the diner. "We believe someone wants to harm you. Maybe your wife too." All this time, she was holding a business card. "Call this person." She handed me a card with only a first name and a phone number on it. "Tell the person you're David Rabbit."

I couldn't help it; I laughed again. "Rabbit? You've got to be kidding. Seriously?"

"He'll make arrangements for you and your family."

"Arrangements?"

"For protection. He's your WITSEC agent."

WITSEC? It took me a moment. Witness protection?

Suddenly, I realized that Paula Merker wasn't really in Hawaii to discuss the legal case at all. She might not be a lawyer with the FAA. She might even be someone who rides around in black SUVs and camouflage helicopters. Whoever she was, one thing was certain—she was serious, and she was here to warn me about a death threat.

CHAPTER NINETEEN

Hazardous Duty

There was nothing subtle about my reaction. My grin changed to a grimace, and my laughter at Paula's spy antics now clogged a mouth too dry for words. Finally, I managed to say, "Why would anybody want to kill me?"

"I don't know all the specifics," Paula said unconvincingly, "but I know our source is reliable."

People I know and cases I had recently worked on ran through my head. My mind was racing, trying to find a reason for this. I began filling in details, making wild assumptions and creating even more outlandish scenarios. None of them seemed to justify somebody wanting to kill me, none of them except possibly Tommy Cho. Was I a part of an elaborate scheme to bust Tommy Cho for something he had done? Did the Feds use me as a pawn to get in to Tommy's operation? If so, the prosecution would have to prove probable cause to make any charges stick. Perhaps I became their probable cause when he pulled a gun on me. But my reasons for entering Island Heli that morning had to be proven valid or the whole house of cards would collapse for them. Of course, the FAA had to ignore that the feds were obviously prepared for a raid, and perhaps they even made the unidentified phone call to the FAA that sent me to Island Heli. The FAA acted as if it didn't know

I was used as bait to bust Cho, but Cho knew, and he wasn't going to ignore it. I was convinced that it was Tommy who was out to kill me. "Are you saying Tommy Cho wants to kill me? And Jill?"

"Where's your wife?" Paula said, ignoring my question.

Her tone was casual, but her question set off all kinds of survival alarms. I didn't know who she truly was or for whom she worked. I also felt certain she wasn't telling me everything she knew. "She's out of town," I said.

"You have to tell me so we can help you." Paula knew that something had just shifted between us. "David, we have to hide you and your family under a new identity."

The idea of Jill living a secret life was laughable. Jill's Italian. There is no way she would give up her family for the rest of her life. "I'll call the number on the card you've given me." I told Paula. "Let me see what I can find out. I have to think about it all." My mind was already at work figuring out how to deal with this crisis myself. One thing I had learned working for the government is that, by design, federal agencies don't put the needs of an individual first. I could never count on them to protect my family. At the time, the paradox of trusting a federal agency to protect air passengers, but not trusting them to protect my family hadn't occurred to me.

When I returned to the office, Peter Undem was curious about why Paula wanted to see me alone. I made up a story about her wanting to grill me more about the details of the Cho case without embarrassing me in front of anybody. Already, my life had changed. Already, I trusted nobody.

When I called Jill that night, I didn't tell her about my meeting with Paula Merker. I didn't want to tell her I'd put our family in danger. Instead, without any mention of death threats or WITSEC, I suggested that before she started looking for work in Colorado, she should visit my sister Theresa in California. If I could keep my family moving around, they would be harder to find. Jill was excited by the trip, and we made the arrangements.

Theresa, my eldest sister, was a successful textiles representative in Los Angeles. She was romantically involved with a man named Gary Crowell*. Although I had never met him, Theresa had told me he was a Los Angeles businessman with some mysterious connection to the National Security Administration (NSA). Earlier, she had asked Gary to help me get transferred, but since no transfer had transpired, I figured he was no more involved with the NSA than I was with the CIA.

I trusted Theresa with my life, but even so, I shared with her only the absolute minimum amount of information about the danger Jill, Tyler, and I faced. During our conversation, I never mentioned Gary, and neither did she.

It was only a few days after that conversation when I received a phone call from David Gilliam, the director of the FAA's Western Pacific region. I had talked to him about my transfer request months earlier, but he had offered no hope of its happening. I knew he was currently away on vacation. "Hi, David," he said cordially, as if we were best buddies "You'll never guess what I was just told." I couldn't even guess why a man close to the top of the FAA organizational chart would be calling me at all. "I don't know who you talked to in Washington, D.C, but you can pack your bags. You're going to Denver."

A few days later, I said aloha to Hawaii, with more sadness than I had expected, and joined Jill and Tyler in California for a vacation. Even so, I was feeling increasingly bad about deceiving my wife. Plus, by declining the WITSEC offer, I had put our fate into my own hands.

One day, unable to handle the pressure any longer, I called the WITSEC number that Paula had given me. I heard a mechanical voice say, "The number you have reached has been disconnected or is no longer in service." I tried again, certain I had dialed incorrectly. I hadn't.

Shaken by the loss of the WITSEC option, I again considered having Theresa approach Gary for his help. Then I realized, if Gary had engineered my transfer, he already knew about the hit; and if he hadn't engineered it, I sure didn't want to babble on about a hit man. I went around and around, looking for a solution before we had to leave for Denver in four days. In the end, I said nothing.

When I showed up for work in Denver, I was given a cubicle that still had photos hanging on the walls. The desk drawers were full of pens and files. I asked the man in the cubicle next to me if I was sharing the space with someone.

"Oh no, that was Bill's office," he said. "He died." I couldn't believe what I was hearing. "That's how you got this job. Someone's gotta die to get on here in Denver."

I realized I was being paranoid, but I went ahead and asked anyway. "He died of natural causes, right?"

"Not really. It was weird." This was not the answer I wanted to hear. "He was coming home from vacation, and he started feeling ill after eating lunch. So he lay down in the backseat, and his wife drove on home. When she got there, he was dead. Heart attack, I guess."

"Well, heart attack's a natural cause," I said, not thinking about how weird I must be sounding. The guy looked at me strangely. Oh shit! I really did need help.

Finally, I decided to turn to the one person whom I could always turn to—my wife. I told her the whole story. When I finished, Jill said, "Don't be ridiculous, Dave. If someone wanted us dead, we'd be dead by now." I knew that in Jill's family, *The Godfather* movie saga was like family history, but "ridiculous" seemed overly dismissive. "This hit man story's a perfect way to get you off the island, even out of the FAA. And have you forgotten about Cho?" she said. "It's nothing but a ploy."

I wasn't totally convinced by Jill, but as the weeks passed without any further mention of death threats, I did stop looking over my shoulder, at least most of the time. I was so busy doing accident investigations, as well as surveillance on repair stations and mechanics' schools, that I had little time to be paranoid. Between November 1992 and March 1993, I was involved in nearly all the thirty-one accident investigations in Colorado and Wyoming. On some, I was the investigator in charge (IIC). On others, I was either the FAA lead investigator or involved in writing or reviewing the accident reports or enforcement actions.

The experience I gained was invaluable. I became very good at my job. My inherent ability to see the actual causes and conditions of an accident—that is, to see Bambi, instead of the driver of the car, as the problem—developed further. Norm Wiemeyer, manager of the NTSB office in Denver, often requested me to assist or consult on his more difficult accident investigations. Over time, we even became good friends. Life was good again.

Years later, I received a call from a woman who identified herself as Stacy Bridger*, and like Paula, she claimed to be an attorney with the FAA. "We're going to trial on the Cho case," she said. "We need you to testify." I didn't say anything. After a long wait, she said, "Can you come to Hawaii, David?"

Sooner or later, I knew I had to get to the bottom of it all. The yoke of fear I had carried around all these years had lessened and felt more and more foolish. Nevertheless, it couldn't continue. I had to know. "Yeah, sure. I can come."

A few days later, I was back in Hawaii. I met with Stacy and several other people, and then the trial began.

Cho's attorney started by arguing that the time-limited, flight-control lever, used to balance the forces on the direction control of the helicopter, did not present an "airworthiness concern." That surprised me. If that was the best defense they had, I knew they would lose.

I sat in the back of the courtroom until Stacy, a petite, young woman who reminded me of a Sally Field character, called me to come forward. As I walked to the witness stand, all eyes were on me. My new black suit, white shirt, and black tie made me look like one of the FBI/DEA/CIA/NSA agents—or whoever they were—sitting behind me in the back row.

Having failed with the airworthiness defense, Cho's attorney decided to attack me. He questioned my character, experience, and qualifications to testify as an expert. He argued that I couldn't act as a material witness and as the technical expert. He then stated that the "FAA could surely find someone more qualified."

After the judge examined my credentials, he pointed out to the court that, for six years, I managed a helicopter-repair facility that was ten times larger than Cho's, managed an aircraft-repair facility supervising more than forty mechanics, and taught aircraft mechanics before becoming an FAA safety inspector. He then added that I had investigated thirty accidents in the past year alone. The judge looked at Cho's attorney and asked who exactly he had in mind who was better qualified.

Throughout all of this, Cho sat and glared at me. The more I watched him, the more I believed Jill was wrong. This man was very dangerous. When the judge ultimately ruled in the FAA's favor, his eyes sparked with hatred.

Great, I thought. *I'm dead.* Well, maybe not. I had come up with my own plan for how to deal with all this, a plan I had not mentioned to anyone. I walked over to Cho and said, "Mr. Cho, I'd like to invite you to join me for coffee at the café across the street."

Stacy shot me a nasty look. "You can't do that."

"Yes, we can," Cho said as he stood up. Three thugs behind him also got up. As we started toward the café, the three thugs and Stacy were right behind us.

"I'd like to talk to you alone," I said, and Cho told his bodyguards to wait outside. I didn't say anything to Stacy, but she stayed with the thugs.

In the café, we ordered coffee and then sat in uncomfortable silence while we waited to be served. I didn't figure I should simply ask Cho about a hit contract, so I tried another approach. I appealed to him as a father.

"I just want you to know, Mr. Cho, that the only reason I came to your hangar that day was just to do my job. No other reason. I'm an FAA agent. Nothing else. And I'm good at what I do. I'm sure you are too."

"I'm the best at what I do." I half-expected him to pull out an Uzi and drop it on the café table to prove it.

"Well, my job's to protect people—not just the ones who fly airplanes and helicopters—but the ones on the ground too. People like my son, who's eight."

"I have a young son, too," he said. I already knew this.

"If you're flying something that's not airworthy—"

"My helicopter was airworthy!"

"It was flight-worthy, sir, but as you know, being airworthy involves more than that." He conceded I was right, so I kept talking. "My only job that day was to protect our two sons, yours and mine. That's what I was doing." Cho never took his eyes off me, but the look he gave me now was different from the cold glare in the courtroom. This time, I was seeing the eyes of a boy's father looking at me. "So what I want to know is, are we okay?"

He gave me a long look before nodding. "You got nothing to worry about from me, Inspector Soucie." He got up from the table. "And you never will. Say hello to your lovely wife and son in Colorado."

"Thank you, Mr. Cho, I will."

He walked off to rejoin his thugs. That's when it hit me. Colorado? How did he know my family was in Colorado? The thoughts I had dismissed as paranoia long ago once again ran through my mind.

Later, I heard that Cho's fine for violating FAA regulations had been reduced. More importantly, I found there were never drug charges placed against him. Somewhere along the way, I realized that whatever angel I had in Washington, Cho had an angel closer to the throne. I don't think he was a drug smuggler at all. I think he was involved with the government, perhaps providing helicopters for undercover government operations. I never knew why Island Heli was raided the day I was there. Maybe Cho worked for the NSA or the CIA and neither the FBI nor the DEA knew this. Government agencies, like airline companies, protect their information at all costs, even from each other. Whatever happened, I never saw Cho again, and no hit man ever appeared.

CHAPTER TWENTY

The Incredible Journey of Tic Tac Man

I returned to Denver a new man, free from danger and free of deceit. Now I could enjoy my wife and my son, without guilt or blame. Once again, I could focus on my goal of improving aviation safety, and I could continue to improve my skills and reputation as an accident investigator.

Before long, I was assigned a case that required all my skills and experience and further convinced me that the fundamental decision-making process itself had to change before aviation would be safer. The day I got the case, I found a note from Walt Wise and the accident investigation field bag on my desk. The note said that I had accident duty that week and that "the new GPS thingy" was in the bag. The FAA had recently added Global Positioning System to the field kit. In 1993, GPS was still new, and I was anxious to use it for mapping an accident scene. As if on cue, the beeper started vibrating, and I called the dispatcher in Seattle.

All the Western Pacific region's communications and coordination were run out of the headquarters there. "We have two aircraft off radar on final approach in Casper," the dispatcher said. "A Mitsubishi MU-2, four SOBs, and a Piper Comanche, two SOBs." *SOB* stood for "souls on board." I loaded up six tox-boxes and headed to Wyoming.

While I drove north into a blizzard, I talked to Norm Wiemeyer, the manager of the NTSB Denver office, on my new Motorola Digital Personal Communicator (cell phone). Norm quickly designated me the IIC (investigator in charge) until the NTSB could send out its man. It was just what I wanted to hear.

To my knowledge, a situation where two aircraft were missing on the same approach path had never happened before. I was prepared for the challenge, even while I was saddened to think about the distinct possibility of casualties. While still on the road, I notified the Casper airport to shut down until the FAA could test the airport navigation equipment. By the time I reached the mobile command center at the Casper airport, the FAA had already verified that the approach equipment was working perfectly. I told the airport manager that he could reopen, but I was no closer to understanding what had happened to the missing planes.

A search-and-rescue team had been on the ground for hours without success. Part of the difficulty was that the ELT (electronic locating transmitter) signal the team picked up seemed to be moving. The rescue team was not aware that a second aircraft had crashed. Once I contacted them by radio and explained that we actually had two missing airplanes, they realized they were chasing their tails by following the two crossing signals. One signal indicated that a plane went off radar at 42.54.34 N latitude, in line with the runway, eight miles out. I sent them to check that location, while I contacted Casper Air Service, the owner of the Mitsubishi MU-2, and interviewed several people. From the interviews, I learned that Wyoming Life Flight had received notification at 12:46 AM, April 6, 1993, of a patient at Riverton, Wyoming, in a state of post-cardiac-arrest care. The patient's name was Jon Willis*. The timeline made no sense to me, so I got in touch with one of the nurses who had cared for Willis.

"It all started when he was driving and threw a Tic Tac into his mouth and choked on it," she said.

"Choked? I thought he had a heart attack."

"Well, yes. But first, he was choking. And that's what brought on a heart attack."

"So you immediately airlifted him to Casper?"

"Well, no. First, we picked him up in an ambulance and brought him here to Riverton."

"An air ambulance, you mean?"

"A regular ambulance, but it slid off the road."

"I see," I said. "Then you put him on the airplane?"

"Well, no. We got another ambulance to bring him here to Riverton, but his heart attack was too severe for us to treat, so he needed to go to Casper."

"On the plane?"

"Well, no. We put him on a helicopter first. Took him out there, loaded him up, and everything."

"What helicopter?" I was getting more confused by the minute. "A helicopter took him to the airplane?"

"Well, no. Actually, the snow grounded the helicopter, so we put him back in the ambulance and took him . . ."

"To Casper?"

"Of course not, to the airplane."

"Why didn't you just send him by ambulance?" I asked.

"'Cause it's a three-hour drive in this weather."

I calculated the time that had elapsed. It was more than twelve hours since the journey started. I thanked her, hung up the phone, and marveled at the bizarre course of events in the life of Tic Tac Man, as I now thought of him.

A local rancher loaned us his Bell JetRanger helicopter for an aerial search. Having learned all I could on the ground, the moment the sky cleared even a little, we took off. We flew a reverse approach from the runway, scouring the snow for signs of the airplanes. All I spotted was one of the two snowcats being used in the ground search. From the air, I could see that the ground search was futile. As soon as the searchers traversed an area, a blowing snow quickly concealed their tracks, leaving them to guess where they had searched and where to go next.

Seven miles out, we turned up a valley to avoid thick clouds ahead of us. After two hours of gray winter sky and blowing snow, we were ready to give up, when we suddenly hit a cloud break and a ray of sunlight flooded the valley below. There, in an endless field of white, I saw a black rock that was larger than any other rock. I asked the pilot to drop down closer.

"We're about out of fuel," the pilot said, "and we have to get back to the airport."

Dark clouds quickly snuffed out the flicker of sunlight, and the snowfall was increasing. If we returned to the airport, I knew I would never be able to find the only thing we had spotted all day. "Just put me down in the valley," I said.

The pilot laughed at my request. "I can't do that. You'll freeze to death. It must be ten below zero."

"I'm a third-generation native of Colorado. I know about mountain cold. Just put me down by the black rock." With my usual overkill, I lectured him about my survival skills and my obligations as lead accident investigator. I didn't bother to point out that "accidental investigator" was a more appropriate title. In any case, the pilot seemed to really look at me for the first time. A position in the FAA comes with a certain respect. He began our descent.

The pilot was afraid of meeting high winds on the valley floor, so he hovered three feet above a ridge. To reach the valley, I'd have to climb down a cliff. "You sure about this?" he said. I opened the door, threw out my field bag, and jumped into a waist-deep snow. While I watched him fly off, I hoped I knew what I was doing.

The wind pushed against me as I climbed down a steep cliff. I paused every few minutes to record my position on the GPS and snap a picture. I made it to the valley floor without killing myself. *That's a start*, I thought. Contrary to the pilot's assessment, the air on the valley floor was completely still. I removed the snowshoes from my field bag, strapped them to my boots, and walked toward the strange, black rock. The snow crunching under my feet was the only sound breaking the eerie silence. I stopped several times to record my position as I trudged across the valley floor. Each time, I scanned the area for signs of an airplane, yet saw nothing but white snow and the strange, black rock.

When I reached the rock and brushed off the snow, I immediately saw that the black color was highlighted with red and purple burned skin stretched drum-tight. The rock was one of the crash victims. "I found one of 'em," I said as I radioed my position to the search team.

"Where'd you find the plane?" the search leader asked.

"I didn't find a plane, just a body." We both knew this meant an in-flight breakup was possible. "Could you get up here fast?" I was ready for a warm snowcat and people to talk to. This was a hard one to figure out.

While I waited, I sat in the snow staring at the body. I tried to imagine what had happened. If the airplane had exploded in the air, pieces could scatter for miles. But I hadn't seen any airplane parts from the air or the ground. I started asking my questions to the accident victim, as I did at every accident scene. Although the dead can't talk, they do provide a lot of information. The body had been dismembered, except for the right arm. The elbow was up in the air and the hand on the ground. It was like he was . . . crawling. That meant that he was alive and in flames when he landed on the valley floor. But crawling where? Why?

As I studied the body for clues, I hardly noticed that the snow had stopped or that the gray clouds had given way to beautiful blue skies. I aligned myself with the direction of what was left of the victim's legs and looked toward the rock wall behind him. About fifty yards away, I saw the white vertical stabilizer sticking out of the snow. It had light blue markings that read "N76LR." It was the back section of the plane. I let my gaze continue out until it reached the cliffs. Then I slowly raised my eyes until I found the front section hanging precariously on the edge.

I snowshoed to the rock wall and then climbed up. From my perch, I could see into the tail of the airplane. The interior was blown to pieces. Clearly, there had been an explosion. Since there was a patient on board, the likely culprit was the two oxygen tanks that fit beneath his gurney. I knew the medical kit intimately. I had designed it. Oh shit!

I *designed it*! I now understood how the black-rock man ended up so far from the accident site. He was the patient. He was Tic Tac Man.

Step-by-step, I pieced together a sequence of events, starting with the large batteries in the tail. When the tail separated from the front section of the aircraft, the wires connected to the batteries shorted out and started a fire that spread under the patient. The fire heated the two oxygen tanks until they exploded inside the tubular structure. The explosion launched the patient out of the airplane tail like a circus stuntman shot out of a cannon. He landed in the valley, his oxygen-saturated clothes and body on fire, and crawled through the snow until . . . Suddenly, the incredible sequence of events struck me. He survived choking on a Tic Tac, a heart attack, an ambulance mishap, a helicopter grounded by a blizzard, a hospital inadequate to treat him, a treacherous flight, an airplane crash, a fall down a fifty-foot cliff when the plane broke apart, and an explosion that landed him fifty yards away. Covered with flames, he still crawled away, only to die right here.

There was a big problem with my scenario, however. Even if it explained the incredible journey of Tic Tac Man, it did not explain what caused the crash in the first place. The MU-2 had broken up and then exploded after hitting the ground, and it had hit the ground 900 feet below the approach glide slope. Until that point, the airplane's approach was normal. The three other bodies were found by the front section. The pilot and the medical technician were still in the plane. The flight nurse was thrown out and lying against a rock as if he were taking in the sun. In addition to the bodies, I found one other thing on top of the cliff that I thought might explain the sudden descent of the airplane prior to reaching the runway—a cell phone.

In 1993, cell phones operated on a frequency that "could" interfere with airplane-instrument readings. I obtained the records for the cell phone that we found on site. It belonged to the flight nurse, Tom Wolf. His wife confirmed that he had called her precisely when the aircraft dropped in altitude. Did his phone call cause the instruments in the cockpit to malfunction, incorrectly telling the pilot to drop lower? When the pilot followed these instructions, the plane hit a cliff. Although the

FCC accepted this theory and used it to cement the restriction against cell phone use during flights, I could not prove beyond a doubt that Wolf's cell phone caused the crash.

The final causal report simply said that the pilot, Tom Rickert, failed to maintain altitude. It was a lame excuse for a cause (a diagnosis). In truth, there is not usually a single cause for an aviation accident, and even the cell phone call that I believe disrupted the instrument readings wasn't the only cause of the accident. I believe the true underlying causes of the crash occurred many months before the accident when the company management chose to use the MU-2 aircraft for medical flights and chose to schedule the pilot to work long hours.

The MU-2 aircraft has a dismal safety record due primarily to its high-performance handling characteristics. It is simply more machine than many pilots can handle without intensive training. Between 1971 and 2008, there were 216 MU-2 accidents, killing 258 people. Had these deaths occurred in one crash, the MU-2 would be infamous. Spread over many flights, the deaths remain mostly unnoticed. Because of its poor safety record, the MU-2 became very inexpensive. Even

The front section of the Mitsubishi MU-2 up on the cliff

Field investigators examine the tail section of the MU-2.

though the aircraft was one of the most difficult to fly, the low cost made it attractive for medical-flight services. Years after this particular investigation, Colorado congressman Thomas Tancredo tried, with my assistance, to push the NTSB and the FAA to have the aircraft grounded. Instead, in an effort to quiet the congressman, the FAA required additional training to fly the MU-2.

A second underlying cause named in the NTSB report was pilot error. Pilot error usually means something happens while in flight, but for Tom Rickert, his fatal pilot error was fatigue. He had already worked a full day in his office when he received the call for the flight. By the time he crashed, he had been working for twenty-four hours.

I pondered what could have been done to predict these events and to prevent them from colliding and causing this tragic accident. Did the owners of the Wyoming air service consider the MU-2's terrible safety record, or did they overlook the dismal safety record in favor of lower cost to purchase? Similarly, did economic conditions pressure the pilot into working as both an administrator during the day and an EMS pilot at night? Once again, it all seemed to come down to corporate decisions. If I wanted to improve aviation safety, this underlying cause had to be addressed, but I had no idea how to do this. Instead, I simply wrote my two reports, one satisfying and one disappointing, and closed the investigation.

The satisfying report covered the second airplane missing in the Wyoming blizzard. We located the plane the same day I found Tic Tac Man. The pilot had run out of fuel and had made an emergency landing in a field. He was a Boy Scout leader who turned the twenty-four-hour ordeal into a survival test for him and his Scout-savvy young son. We found them both alive and well. Unlike my Tic Tac Man report, the report I filed on the Boy Scout accident was brief and not at all mysterious. It seems good endings are always this way.

CHAPTER TWENTY-ONE

The Flight of Practical Cats

People whose jobs bring them face-to-face with death, especially a gruesome death, learn to distance themselves from what they do and see, often by developing a dark sense of humor. If I tell the story of the incredible journey of Jon Willis, Tic Tac Man, to almost any homicide detective, emergency medical technician, fireman, ER nurse, or aviation-disaster investigator, they laugh—not because death is funny, but because it's how we all cope. Most of us, if we pondered the fragile and fickle nature of existence too long, would become incapable of functioning. Perhaps the awareness of life's fragile membrane had surfaced and had refused to be laughed away, or perhaps there were other reasons, but as 1993 was coming to an end, I daily scanned the FAA journals and weekly updates for job opportunities, temporary assignments, and other duties as assigned (ODA), looking for an opportunity that would give me a break from death duty. Once again, Walt Wise came to my rescue.

One day in mid-January, I was spinning around in my squeaky chair, lost in my thoughts, when Walt came up to me. "Earth to Dave!" he said. "You in there?"

I spun around. "Fuck you, weekend warrior. Don't you have anything better to do than bother me?" Walt was in the National Guard, and I teased him about playing soldier.

"Oh, okay, asshole. I'll just leave you alone. You probably wouldn't want to go to London anyway."

I jumped from my chair and grabbed at the paper he was holding. "Give me that thing."

He threw the paper at me and stood there smiling. In my haste, I had knocked over the coffee dregs in my cup and had to dab at them while trying to read the wrinkled assignment request. The Eastern regional headquarters had an assignment in London that required maintenance inspectors with extensive repair-station backgrounds.

"I thought of you," Walt said, "but if you're busy . . ."

"Jill's been wanting us to go to London for years," I said. "She really wants to see *Cats* over there."

"Sounds purrrfect," Walt said, unable to resist.

I didn't waste any time applying for the assignment, and thanks to my pre-FAA experience as a manager of several repair stations, I was selected along with three other inspectors from around the country. Within two weeks, Jill and I were on our way to England. As we boarded the flight, I still had no idea what my actual assignment was. I was told I would be briefed when I arrived.

We bought a ticket to London for Jill, but I was catching a free ride by doing an en route inspection, that is, an in-flight surveillance performed in the cockpit. I was to examine everything that went on during the flight to see if there were any violations of FAA regulations or approved procedures. I had done these types of inspections many times before. These inspections were required by the FAA, and Peter Undem assigned me responsibility for them since it provided me with a way to get home occasionally to see Jill and Tyler in Colorado. Naturally, the crew hated these inspections. It was like having a cop riding in the backseat while you drive cross country. I always tried to cut the tension with small talk and friendliness, but everyone knew that with a click of my pen, the airline would face a ten-thousand-dollar fine. The check marks on the Program Tracking and Reporting System (PTRS) could add up quickly.

After my walk-around with the pilot and the routine preflight briefing with the crew was complete, I checked the pilot's certificate and the Flight Operations Manual to verify they were current. Then, the pilot gave the command to push back from the gate. As we did, I noticed that the split-flap-indicator warning light was on. I asked him about it. "Oh, that was flickering on when we landed," he said. "The flaps are fine. I'm sure it's just an indication problem."

The pilot might have been sure, but I wasn't. "Is that right?" I said. "May I see your maintenance log?"

The pilot knew he was busted. He knew his log would show that he hadn't had a mechanic look at the problem at all. Instead of providing the log, he called to have the Jetway brought back out, and then he called maintenance.

Once we docked with the Jetway, the door opened, and a large man wearing an orange jacket that was two sizes too small came into the cockpit. He glared at me. Nobody was pleased. The aircraft was now delayed. Delays cost money and upset the passengers, and I was the cause. I stepped out of the cockpit into the galley to let things cool down a bit while they addressed the problem. I could see Jill a few rows back. She gave me an eye roll that said, "If I miss *Cats*, I will kick your ass."

A few minutes later, the mechanic came out of the cockpit and handed me the logbook. "Here ya go! Happy now?" He walked off the plane and closed the door.

I took the logbook with me into the cockpit and buckled my seat belt once again. The logbook entry surprised me. It said, "Cycled the flaps and could not duplicate the problem." The aircraft pushed back again.

The split-flap-indicator warning light was no longer lit, but something still felt wrong to me. "Could you push the PTT button for me?" I asked. The PTT (push to test) button turns on all the indicator lights to confirm that they're working. The pilot pretended not to hear me, but the copilot pushed the PTT button. Every light in the console lit up except the split-flap-indicator warning light. I was sure that the mechanic had come on board and simply removed the one lightbulb, but I played dumb and said something inane like, "Oh, look at that.

The bulb must have burned out. We better get that mechanic up here to change it."

Several mechanics came on board this time, and after conferring with the pilot, they went down to look under the right wing of the airplane. I followed them down the stairs onto the ramp and watched them remove the flap actuator on the right wing. It was damaged and out of sync with the left-flap actuator. Without proper sync between the two, there was a high probability that the flaps would have retracted at different speeds after takeoff. This could have sent the airplane into a severe and unrecoverable roll, leading to a crash. The pilot knew the flap-indicator light was not functioning properly, but it didn't register with him that it could lead to a crash. Why? Because he had flown the 747 to England so many times without incident, it was routine.

The pilot was like a driver who takes the same road to work every day without incident, until one day a stop sign is placed at a new intersection. He sees the sign, but the change doesn't register. He drives through it. His routine is too familiar for him to recognize the change. Does he crash? Maybe. Drive safely to work? Maybe.

I could not prove that the 747 leaving New York for London would have crashed if I had done nothing. That's what makes the job so difficult. FAA inspectors never know whether grounding a plane prevents an accident. We never know when we're right, only when we're wrong, only when we fail to take action and a disaster occurs. Einstein said, "God doesn't play dice." Neither do I. I wasn't willing to gamble.

The proper repair took more than four hours, and by the time it was done, the pilot had cancelled the flight. We were all off-loaded and spent the night in a New York City hotel. We left for London the next morning, but by the time we arrived, Jill was unable to get replacement tickets for the play. To this day, she insists that she missed seeing *Cats* in London because of a burned-out lightbulb.

Once we arrived, I was told why we were there: Lucas Industries. Based in Birmingham, England, Lucas is a well-known manufacturer of auto and aerospace parts. The division of interest to us, Lucas Aerospace, manufactured many parts for the 747, including the actuators, the very part that had malfunctioned on our flight. The London Lucas facility was approved by the FAA only to manufacture parts. They were not an FAA-approved "foreign repair station," so they were not approved to repair or overhaul them. When parts required repair or overhaul, the airlines had to send them to a Lucas-owned, FAA-approved repair station in New York.

The FAA had learned months earlier that when the New York station had too many parts to repair, they sometimes sent them to the London manufacturing facility for repairs. This practice, although acceptable in Europe, was a serious violation of FAA regulations. In the States, aviation regulations are found in the Federal Aviation Regulations (FAR) within the U.S. Code of Federal Regulations (CFR). In Europe, airlines and repair stations are guided by the Joint Aviation Regulations (JAR), which allows for repairs of parts under different conditions that the FAR does not.

The FAA did not impose fines or sanctions against Lucas for violating the FARs, nor did it require immediate replacement of all parts that were improperly repaired. Instead, we were instructed to work with Lucas on "a quiet solution to the problem." Since more than half the 747s might have had illegally overhauled parts installed on them, replacing them would have had a devastating economic impact on the aviation industry. Furthermore, had the FAA fined the airline and grounded the fleet, it would have been an admission that it had knowingly compromised public safety by not immediately responding to the problem months earlier. Another inspector told me that the phrase "a quiet solution" was a code for covert operations. It meant "Don't let the public know what we're doing. Don't alarm anyone."

The FAA divided up Lucas' twelve aerospace facilities among the four of us. We were told to visit our assigned locations and do whatever was needed to certify them as "foreign repair stations," all within three weeks. Considering that the certification process typically takes between three and six months, it was an impossible task, if done properly. Our objections and arguments to the New York FAA international office and FAA headquarters in D.C. fell on deaf ears. Nobody at the FAA told us directly to simply approve the repair facilities using the lesser JAR standards, but we all understood that this action would be an acceptable solution. To do so would transfer the problem from the FAA to the inspectors if anything went wrong. I envisioned the headlines: "FAA Inspection Team in London Cuts Corners and Risks Passengers' Lives!"

I suspected that some of the FAA inspectors had approved the repair stations to which they were assigned because they finished long before I did, but I refused to do that. Instead, I required the Lucas facilities to satisfy, step-by-step, every FAR for a foreign repair station.

After three weeks, I returned to Colorado and awaited the follow-up assignment to implement the changes at Lucas Aerospace. Maybe Jill would still see *Cats* in London. Maybe I'd be off the hook. When the implementation team was announced, the three other inspectors were on it, but I was not.

What I did wrong was insist that FAA regulations be implemented. What I did wrong was ignore the reason we were sent on a covert operation—the FAA didn't want to risk upsetting the public. Maximizing aviation safety didn't seem to be a high priority of the FAA on this operation. The FAA has two mandates. One of the mandates is to regulate aviation and to ensure safety, and the other mandate is to promote aviation. I was getting my knuckles rapped as a reminder of which mandate mattered most to the FAA.

While I nursed my bruised ego, life at the Denver FAA office again settled into the routine of inspection and accident investigation. Both the FAA and the aviation industry in the United States were in for a bumpy ride in 1994, but the worst hadn't hit yet. That July, the conference room at the Denver Flight Standards District Office (FSDO) weekly "stand-up" meeting was packed. The intent of the meeting was a brief knowledge sharing of events. In fact, the meetings were rarely brief, and little knowledge was shared. Dan Shelton, the office manager, chaired the meeting. He was a distinguished-looking, tall, black man with a calm demeanor and an infectious laugh. He typically skipped hellos and simply opened the meetings with a loud pronouncement. "Somebody here has to take a facilitator course," he bellowed to more than fifty inspectors. No one knew what a facilitator was, but we did know that volunteering for anything new in the FAA was painting a target on your back.

The room got totally still. We all cast our eyes to the ground, trying not to call attention to ourselves. Shelton paced the room, carefully scanning for even the slightest eye contact with someone. Anyone would do. Instinctively, I slid down in my chair like a schoolboy praying not to be called on to answer a question, but as I did, my government-issue steno pad fell from my lap and slapped the floor loudly. The entire room looked at me with relief. "That's great, Soucie," Shelton said. "You'll be great at this facilitator thing. You love the limelight, and people like you. You leave tomorrow for Seattle."

CHAPTER TWENTY-TWO

The Puzzle Master

In 1994, all federal agencies were examining their operations as part of President Clinton's plan for a more efficient government. One way the FAA chose to achieve this goal was to improve the effectiveness of its meetings by training facilitators to lead them. The facilitator's job was to steer the meeting when the group lost their way and to offer different approaches to resolving internal challenges. To do this job, the facilitator would receive special training and be armed with a set of tools.

I attended my first facilitator class in July 1994 in a makeshift classroom in a downtown Seattle hotel. I was given a canvas bag that held some reference books and dry-erase markers. My instructor called it my "facilitator toolbox." The class taught us "soft" tools to deal with personalities and personal conflicts and "hard" tools for organization, planning, and general management. I quickly learned how to use all my new tools skillfully. Dan Shelton had been right to select me for this class.

Between 1994 and 1996, I spent as much time as a facilitator as I spent investigating accidents and performing surveillance on repair stations and airlines. There were no commercial-airline accidents, only private- and small-airplane accidents. Even so, I investigated enough aviation disasters to remain overly familiar with carting around tox-boxes, and I still could never load them in my vehicle without a sense of sadness for what awaited me. The death of a few people in a private- or corporate-airplane crash, while statistically unimpressive, nevertheless changes the course of life for many people, usually the wives and children left without husbands and fathers, sometimes the employees of a business that folded without its owner, or sometimes the friends who lost the last good person they could count on in life. This awareness stuck to me as surely as the smell of death stuck to my skin and my hair, no matter how many times I would shower, or how hot the water, or how much I scrubbed.

My daily duties, when I was not being a facilitator, involved more facility inspections and routine paperwork than they did accident investigation. Yet, it is the unusual events that occur, whether they are tragic or simply unbelievable, that we remember.

There were 280 aviation disasters that our Colorado-based FAA office investigated from 1994 to 1996. Fifty-three of the accidents led to ninety-six deaths. On many of these accidents, either I was the lead investigator or I assisted with the NTSB investigation. Some accidents I investigated are remembered because they defy belief—like the pilot of a Piper Comanche who mistook the blue lights on a high-tension power line near Limon, Colorado, for runway lights and landed on them. Only one passenger was on board, and once I saw how the airplane had been turned into a lump of molten aluminum from the electrical fire, I knew I would need the two tox-boxes I had brought. Instead, both men survived.

Other investigations stick with you because the situations were so bizarre, like the day I was called out to Fort Lupton, Colorado, to investigate the death of a parachute jumper in a farm field. In truth, the farmer, also a pilot, was running an illegal jump operation. On that day, he had five people on board. After they jumped, he landed the plane, raced to where the guy whose chute failed to open had hit the ground, threw his body in the back of his pickup, parked the pickup in his hangar, and then went out in another truck and picked up everyone else. They all expressed gratitude, paid him, and left. Afterward, the farmer reported that he had found an unknown jumper dead in his field. Once I learned the truth, we shut him down with a federal judge's order; six months later, he was back in business. The second time I went out with federal marshals, they arrested him, and he went to jail.

Occasionally, you are fortunate enough to investigate something that is actually humorous. This was the case when the buyer of a small plane reported that the seller, a well-known investigative-television reporter in Denver, had, prior to the sale, replaced the airplane's failing alternator with a much less expensive, but temporarily functional, automotive alternator. The television reporter was famous for shutting down a restaurant after spotting a cockroach or for uncovering a crooked roofing company or such, so he was very upset when I wrote him a violation. I admit to having enjoyed telling the story at parties, but the FAA being what it is, the last laugh was on me. My FAA boss decided he didn't want the news reporter retaliating by digging up an incriminating story about the FAA, so he waived the violation and told me to forget about it.

Naturally, there are investigations that you truly wish you could forget, but they are permanently cut into your heart, like initials carved in the bark of an aspen tree. No matter how many years pass, that cut remains. For me, the day I wish I could forget was the day I was sent on a four-tox-box (4TB) call to Afton, Wyoming. The small plane remained on the runway, and the four people were still inside—father, mother, and two children. The father was the pilot, and his hands were on the wheel. All four were strapped in their seats for takeoff. The father had forgotten to open the flow of fuel, and after takeoff, the plane quickly stalled and dropped twenty-five feet back to the runway, where it bounced up and down until it finally settled where I found it. Everyone inside might have survived, but the bouncing or sudden drop caused their hearts to explode. When I found them, looking normal except for their eyes, for all their eyeballs had popped out and were hanging, I could not believe they had died. The mark this accident made on my heart will always be part of me.

Oddly, despite all the death experiences, the investigation that haunted me the most involved finding no bodies at all. Everyone on board had simply vaporized. It was June 21, 1996, when Northwest Mountain Region Dispatch contacted me, saying, "We have a Beech

58, a Baron missing. Douglas County Sheriff's Department is looking for the aircraft. Four SOBs." I hung up the phone and got four tox-boxes from the closet.

As I approached the accident site near Colorado Springs, I saw the flashing lights of the Douglas County sheriff's car through dense fog and a hazy setting sun. The car was parked on the side of I-25. The deputy recognized the FAA Jeep and stepped up to my window. "I think we found it up there," he said, pointing to the yellow crime-scene tape on the hill above us.

"You found the plane?" I asked.

"Well, there isn't much there," he said.

I didn't want to lose the remaining light, so I shifted into four-wheel drive and headed up the hill. On the top, the fog was thicker, and I nearly drove right through the scatter point. It was the largest scatter point I had ever seen. A fifteen-foot scar that began on a slope was followed by a thirty-foot gouge three feet deep. I stepped out into the fog and immediately felt something different. This was not a typical accident site.

There was no wind, and no birds were chirping. In total silence, I walked around the huge rupture in the earth, shining my flashlight; some metal shone back at me. In the hole, I found a propeller spinner, propeller rings, and one of the three propeller blades. The gouge curved west, where a path of small metal pieces extended through a stand of scrub oak thirty feet long. In the stand, I found part of the left wing tip and the left aileron. Beyond them, in an open area, pieces of the forward fuselage and wings were scattered over a three-hundred-foot path of destruction. The left engine was in some trees 283 feet away. The leaves and other vegetation were scorched and burned. This was the only evidence of fire that I found.

I usually started an investigation at the scatter point and then paced off the distance to the first body part I could find. The FAA had a formula for determining the scatter distance, but it was ineffective. The formula limited the variables to aircraft speed and altitude and neglected the projectile's weight, size, drag coefficient, angle of separation, wind, and other effects—all of these dramatically changed the travel and velocity after impact. I perfected my own algorithms for determining the speed and angle at which the aircraft hit the ground. In addition, using my formula, I could determine if aircraft controls were functioning at impact, if the pilot was attempting a recovery, and if the aircraft engines were producing full power, were throttled back, or were not running at all. This was part of the scientific side of accident investigation, and it worked great up to a certain point.

In truth, aviation-disaster investigation is often more art than science. The investigator is like someone trying to put together a puzzle while wearing a blindfold. As such, I often relied on the bodies of those who were on board to tell me what had happened. I knew that if the arms came off, the impact speed was over 180 miles per hour. At 220 miles per hour, the legs come off. I even devised a formula to calculate the impact angle of the aircraft based on where we found the heads of the pilot or passengers.

While analyzing the spread of human tissue and the location of body parts and metal shrapnel often told me what happened, at this crash site, I walked over three hundred feet from the impact point, but there was no evidence of the pilot or passengers. I sat on a small rock to gather my thoughts. "Talk to me, guys," I said. I always listened to what the dead could tell me. I knew that most investigators did not expect the dead to share their secrets and that some of my cohorts saw me as a mix of the horse whisperer and Edgar Cayce, but I had spent half of my life married to a woman with psychic intuition, so I never doubted my approach. The dead had always told me something. I sat and waited, and waited some more, but for the first time ever, I felt nothing and understood nothing.

A light breeze now rustled the aspen leaves. The fog was beginning to lift. I looked up and saw the late-evening sun painting a beautiful red against the blue sky. Then something caught my eye, something in one of the aspen trees. I went over to examine it and saw it was a piece of flannel cloth. I picked it off the tree limb with a branch and held the material tightly in my hand. "There you are," I said to the shadows. "There you are."

There were actually two pieces of cloth each about six square inches held together by a button. Earlier that day, a man had picked that shirt from his closet, put it on, and buttoned it up with every expectation of unbuttoning it again that night. Often death comes without warning. The realization of how life's plans can be instantly abducted by death sent a chill up my spine.

As I looked through the trees, I found more pieces of the shirt. I soon realized that the trees were also adorned with small bits of human flesh and blood. I became faint and sat down in the leaves and grass, still holding the piece of shirt in my hand. Next to me, I saw a flyer that read, "Promise Keepers 1996, 'The Power of a Promise Kept.'" The Promise Keepers is a Christian organization founded in 1990 by then-coach of the University of Colorado football team, Bill McCartney, and Dr. Dave Wardell. I knew a Promise Keepers' convention was being held in Boulder the next day. I later learned that all four men who died in the crash were on their way to the rally. Two of them were to be speakers.

It took a while, but I eventually pieced together the Promise Keepers' puzzle. The flight originated in Kansas. On their way to Boulder, they encountered a huge thunderstorm over Monument Pass. Air traffic control contacted them and asked them to divert around the storm.

"No, I think I see a hole through it," the pilot said. Ignoring the warning, he flew into the storm.

The plane had flown close to Cheyenne Mountain, which is under surveillance due to its military importance. The military provided us with radar pictures of the plane prior to the accident, flying at a normal 180 to 200 miles per hour. By extrapolating the distance and time between the last photo and the impact point, I calculated its speed to be nearly 700 miles per hour. It was being powered by the storm, and not the engines, at a speed far beyond its structural capabilities. At this speed, the impact put so much pressure on the metal that the fuselage broke into small pieces, and everything and everybody in the plane literally vaporized.

Even though I had figured out the cause of the Promise Keepers' accident, as I had so many others, I was no closer to finding the root cause of aviation accidents or determining how to predict and prevent the next accident. I knew I was a good accident investigator. I had developed an ability to see a disaster from the viewpoint of the victim. I could figure out the puzzle. But so what? It didn't help the dead. I was merely reacting to disaster, not preventing it. If I wanted to improve safety, I realized I needed to develop a new strategy, a new way of thinking about cause and effect, a way of answering the questions before the crash, not afterward. The old way of thinking was not going to save anyone's life.

I reexamined my investigative approach. There had to be something that my years as a mechanic, executive, FAA inspector, and accident investigator had taught me that I could use to predict and prevent accidents. I talked to colleagues. I searched for new points of view in books. No matter how hard I searched for a new approach, I was stumped, until I started to include what I had learned facilitating meetings over the past two years. At first it seemed unrelated, but recalling my experience as a facilitator gave me the new perspective for which I had been searching. It had allowed me to look beyond the issues at hand and see the way cultures, diversity, and environmental factors impacted decision making. In my role as observer, I also saw that the decision-making process in the FAA and in the aviation industry was riddled with problems.

That was when Newton's apple fell on my head, my aha moment, when I knew, without a doubt, that decision making at the corporate level, both in industry and in regulatory agencies, was involved in almost every accident I had investigated. I flashed back to the MGM Grand fire—same thing. Corporate decision making was my one certainty, so it became my starting point, my scatter point. To improve aviation safety, I would have to go to where the decision-making process began: to Washington, D.C., to aviation boardrooms, and to the various agency offices. Even though I had no idea how to get there or what I would do once I arrived, now at least I saw the direction to go. I had no idea that the misdeeds of the FAA itself would provide me with both clarity and opportunity.

CHAPTER TWENTY-THREE

License to Kill

The year 1994 was a bad one for aviation in the United States, with more than one thousand deaths in over two thousand civilian-aviation accidents. Every year, the majority of fatalities, and sometimes all of them, involve accidents in small private planes, air taxis, and commuter flights; but this year also saw 237 fatalities in larger commercial-aviation accidents. During one four-month period, there was a flurry of disasters: a US Air flight crashed in Charlotte, North Carolina, killing 37 passengers; another US Air crash near Aliquippa, Pennsylvania, killed all 132 people on board; and American Eagle Flight 4184 crashed near Roselawn, Indiana, killing all 68 passengers and crew. The public wanted answers, and they expected the FAA to provide them. Instead, the FAA itself would be targeted as the main cause of the crashes.

A warning about the firestorm that hit in 1994 had been sounded a year earlier by FAA employee Mary Rose Diefenderfer, the principal operations inspector (POI) assigned to Alaska Airlines, when she uncovered a web of lies and deceit inside the Seattle-based airline. Diefenderfer reported that Alaska Airlines' pilots were falsifying training records to show that they were trained to fly certain airplanes when, in fact, they were not, putting passengers at risk.

After Diefenderfer blew the whistle on Alaska Airlines, her direct supervisor, the FAA Northwest Mountain regional division manager, removed her from her POI position and reassigned her. She filed a protest. Once this happened, FAA security began an investigation. By the time it was over, five management pilots at Alaska Airlines had lost their commercial-pilot certificates and a senior vice president had lost position. Mary Rose Diefenderfer won her old job back.

I knew Mary Rose and had followed the story closely. Her actions would reverberate in my own life in 2002. I was assigned to the Certification Process Study (CPS) response team, which investigated the possible systemic causes of an Alaska Airlines' crash on January 30, 2000, in the Pacific Ocean a few miles north of Anacapa Island, California. All five crew members and eighty-three passengers were killed.

Years earlier, the FAA had chosen not to punish Alaska Airlines for its violations but, instead, had punished its own employee who had discovered and revealed the violations. As such, Diefenderfer uncovered not only a criminal behavior within Alaska Airlines but also a pattern of negligence, mismanagement, and abuse in the FAA. It was only the beginning.

In December 1994, the *Los Angeles Times* ran a front-page article by Jeff Brazil regarding a four-month study that exposed the faults of the FAA. Mr. Brazil wrote:

> The FAA, the federal agency responsible by law for ensuring air safety, has been named by the NTSB as a cause or factor in 103 airplane accidents and incidents between 1983 and last July [1994] that together killed 574 and injured 421. And records also show that hundreds more people have died in crashes caused by problems to which the FAA had been alerted but failed to act.

The article sent waves of panic and shock throughout the agency. Copies of the article traveled through every office: "FAA Under Fire for Its Slow Response to Safety Warnings." In addition to runway collisions, Mr. Brazil listed other problems of which the FAA was aware but on which it failed to act, including turbulence caused by the wings of Boeing 757 jetliners, procedures for de-icing wings before takeoff, the refurbishment of aging aircraft, passenger access to emergency exits, and the installation of devices that warn if an airplane is flying too low. Perhaps even more damaging, Brazil had a long list of errors made by air traffic controllers and others supervised by the FAA that had directly led to aviation crashes and the loss of life. In the three crashes noted earlier—North Carolina, Pennsylvania, and Indiana—the FAA was directly cited as being responsible. The *Los Angeles Times* story quickly gained momentum when other media outlets picked it up, each of them adding something new to the list of sins of the FAA.

To avoid association with all the negative press, congressmen and senators on the aviation subcommittees began pointing fingers and questioning the leadership of the FAA. The administrator of the FAA at this time was David Hinson. I had met Hinson some years earlier in Chicago when he was the head of Midway Airlines and I was in charge of the Omni Repair Station. I was not a fan of David Hinson. I suspected he was capable of pulling a Roy Morgan if profits were being squeezed by ethics. I had been told Hinson had betrayed Executive Helicopters the same way Roy Morgan had betrayed SFENA after selling them the rights to manufacture the med kit we invented at Air Methods.

The owner of Executive Helicopters told me he had signed a contract with Midway Airlines to provide transport for its passengers from Midway Airport to downtown Chicago and to selected hotels. With a contract negotiated by a Midway vice president in hand, Executive purchased two new helicopters and hired eight pilots and five mechanics, many of whom moved their families from long distances. When Executive's owner went to Midway to pick up the promised deposit check of five hundred thousand dollars, the deal-making vice president was no longer employed there, and Hinson claimed that since he hadn't signed the contract himself, it was invalid.

Because of this, I did not trust David Hinson, so I had no doubt that he would find ways to deflect the criticism directed at the FAA and himself. The first step for the FAA was to publicly examine itself, a visible mea culpa. Because of this, the groups I now led as facilitator were charged, more often than not, with examining how the FAA worked and how to improve it. I realized at the time that this effort was more image repair than genuine reform.

Aviation disasters did not decrease in 1995, and neither did the criticism of the FAA. By 1996, the FAA was desperately in need of good publicity. Seven-year-old Jessica Dubroff promised to provide that and more.

Jessica was a personable young girl passionate about flying. When she announced on television that she would attempt to beat the world record for the youngest person ever to fly coast-to-coast, it seemed as though the aviation gods finally had smiled on the FAA. The record was held by a California boy, Tony Aliengena, who was nine years old in 1988 when he flew across the country and back, accompanied by his flight instructor. Both Jessica's flight instructor and her father, himself a pilot, would accompany her. She planned to make the journey in fewer than five days.

Everyone, including me, saw the incredible potential this flight had for promoting aviation. Even those of us who thought the FAA deserved a good butt-kicking had grown weary of having sore butts. If anyone could boost confidence in flying again, it was surely this little girl. It seems most of us simply ignored the fact that this little girl was years away from even being able to qualify for a pilot's license. We were desperate.

The FAA took every opportunity to tout the event and to crown Jessica another Amelia Earhart. As she got ready to depart from Half Moon Bay, California, one headline read, "From Sea to Shining Sea."

My head was filled with thoughts of my son, Tyler, age twelve, behind the wheel of an airplane. I even called Jill to make sure that she and Tyler were watching the news coverage of Jessica's flight. I had no idea I would regret this call for many years.

Jessica's first stop would be Cheyenne, Wyoming. After Cheyenne, she was headed to Lincoln, Nebraska. Her final stop would be in Washington, D.C.; she had invited President Clinton to join her there on a brief flight.

The journey from California to Wyoming went smoothly, and Jessica, Flight Instructor Joe Reid, and her father, Lloyd Dubroff, prepared to leave early the following morning for Nebraska. Several network news crews swarmed the Cheyenne airport to do interviews and catch a glimpse of the city mayor wishing Jessica luck, while in Denver, most of the FAA office gathered around the television in the conference room to watch. The two men were interviewed, as was Jessica's mother, but it was Jessica who stole the show with her bubbly personality. Her enthusiasm for aviation and her innocent charm blinded us to the rain, wind, and lightning and thunder in the background. We saw only a bright, sunny day in the history of aviation.

When the interviews were over, we turned off the television and returned to our cubicles with renewed enthusiasm and a sense of relief. Things were looking up.

A few minutes later, I heard the accident-duty beeper sound off in the adjacent cubicle. Jeff Graves, a fellow inspector, was on accident duty. I heard him pick up the phone and call the regional dispatch center in Seattle. I also heard him say, "Oh my god!" Jeff's voice sent chills down my spine. "Somebody grab three tox-boxes and bring them to the door while I get the Jeep," he yelled.

"What's going on?" I asked Jeff.

"She crashed! Jessica crashed, and I'm heading to Cheyenne," he replied in a stunned voice.

The Cessna 177 Cardinal crashed just after takeoff in Cheyenne, Wyoming, killing all three people on board. My first thought was for them and their families. My second thought was less noble. I was thankful I wasn't the lead investigator that day. I knew this one was going to get ugly. But in the end, my curiosity propelled me to get into my own car and follow Jeff to Wyoming, even though I was not officially assigned to work the disaster.

It didn't take Jeff long to focus his investigation on the weight of the airplane. He strongly suspected that Jessica's father had filled the Cessna Cardinal with too much equipment, and in the bad weather, the airplane, which was more suited for lower-altitude eastern flying than higher-altitude western flying had lacked enough power to clear the mountains. To confirm his suspicions, he off-loaded every piece of Jessica's belongings onto a scale right in front of the news cameras.

Jeff Graves was right about the airplane, and I was right about the ramifications. The FAA, previously a strong supporter of the flight, unleashed a media barrage of criticism directed at "those who encouraged Jessica to compete for a world record." In other words, the FAA public relations department did what it had become good at doing—spinning the truth to suit its purposes. Its purpose at the time, for the most part, was to reassure the public that everything was fine. Five days after the crash, FAA administrator David Hinson addressed Congress without ever once mentioning Jessica Dubroff. It was disheartening to witness. It was disheartening to think that not even the death of a bubbly young girl could get the FAA to change. It would take another tragic disaster one month later for that to happen.

CHAPTER TWENTY-FOUR

The Summer of Deadly Lies

All I had at this point was a belief that the underlying cause of most accidents began with a flawed decision-making process, so I tried to verify my belief by reviewing many of the accidents I had investigated from a new perspective. As an accident investigator, my job was always to figure out what happened—what caused the crash and how it happened. But if the conditions leading to an accident were already in place long before I arrived at the crash site, I needed to look upward on the chain of events toward the decision makers instead of down the chain to the crash victims. I needed to start looking for the *why*.

I had my pick of accidents to study: Mike Myers, the Promise Keepers, Tic Tac Man, Sergeant Boyd, C. B. Lansing, and on and on. My file cabinet was a graveyard of human error. All accidents, it seems to me, are human error, unless you blame God, nature, karma, fate, magic, or whatever, so the basic questions are: where was the true underlying error made and who made it?

Every spare moment I had was spent reviewing these accidents. After two years of public and congressional criticism of the FAA and my own sense of betrayal by its leadership, I needed to restore my belief that I could somehow improve aviation safety. Maybe by doing so, I also would restore my belief in the FAA. It never occurred to me that the FAA, from top to bottom, was about to implode.

It was May 11, 1996, and 105 people, some on their way for Mother's Day visits, boarded ValuJet Flight 592 in Miami bound for Atlanta. ValuJet, a discount airline in business only three years, was highly successful because of its low prices. After a delay of more than an hour, the twenty-seven-year-old DC-9, previously owned by Delta Air Lines, took off at 2:04 PM and began a normal climb to cruising altitude. Nine minutes later, it plunged into the Florida Everglades.

Investigators learned that, right before takeoff, chemical oxygen generators had been placed in the cargo compartment in five boxes marked COMAT (company-owned material) by ValuJet's maintenance contractor, SabreTech, in violation of FAA regulations forbidding the transport of hazardous materials in aircraft cargo holds. A cargo manifest listed the expired canisters as being empty when they were not. Investigators theorized that the plane experienced a slight jolt while it taxied onto the runway and an oxygen canister activated, producing oxygen and heat, which ignited a fire. Oxygen from other generators fed the resulting fire in the cargo hold. This caused a spare semi-inflated aircraft wheel in the cargo hold to explode.

ValuJet preparing for takeoff

Searchers found no survivors in the Florida Everglades.

Smoke and fire in the passenger cabin were reported six minutes after the plane left Miami International Airport. As soon as the fire was discovered, a flight attendant entered the cockpit—the intercom system to reach the cockpit was not working—and advised the flight crew. Passengers' shouts of "Fire!" could be heard on the cockpit voice recorder during the time the cockpit door was open. The crew immediately asked air traffic control for a return to Miami, and Captain Candi Kubeck and First Officer Richard Hazen were given instructions to return. By then the blaze had already burned through the jet's flight-control cables that help steer the aircraft. A minute later, First Officer Hazen changed the request from Miami International to the nearest available airport.

Three minutes after the fire was discovered, the ValuJet nose-dived. It crashed in the Browns Farm Wildlife Management Area in the Everglades at a speed in excess of five hundred miles per hour.

Kubeck, Hazen, the three flight attendants, and all 105 passengers aboard were killed.

A fisherman watched the airplane fall out of the sky and called 911. Emergency crews arrived but had to park a mile away and use boats to search the deep-water swamp for possible survivors and wreckage. In addition to the isolated location, saw grass, alligators, and the risk of bacterial infection made the recovery effort extremely difficult.

A person is often defined by how he responds to a crisis. The days following September 11, 2001, defined the entire career of New York City mayor Rudy Giuliani, while the single statement, "I'm in charge," made by Alexander Haig, President Reagan's secretary of state, after Reagan was shot, overshadowed Haig's brilliant military and political career. Crises can also define agencies and companies, and both the FAA and the Department of Transportation (DOT), the cabinet department that oversees the FAA, handled the ValuJet crisis like President Nixon handled Watergate. In the process, they damaged their agencies' credibility with the public and destroyed the morale of their employees.

Both Federico Peña, the secretary of transportation, and David Hinson, the FAA administrator, rushed to Florida and proclaimed ValuJet to be safe. ValuJet has "in some cases even exceeded the safety standards that we have at the FAA," Peña said at a May 12 briefing, less than twenty-four hours after the crash. Hinson also assured the public that ValuJet was safe to fly. To prove their point, they boarded a ValuJet flight.

A month later, on June 16, 1996, the FAA grounded ValuJet as unsafe. It was allowed to fly again on September 30, 1996, but ValuJet never recovered from the crash. In 1997, it merged with AirTran Airways. The name *ValuJet*, by then, was too poisonous to continue using it.

It didn't take long for ValuJet's poor safety record to surface. Once it did, many families of the Flight 592 victims were outraged that the airline had not been prosecuted. Despite what Secretary Peña told the public, ValuJet had a much higher accident rate than the ten largest airlines. Even worse, records indicate that both Secretary Peña and David Hinson knew, or should have known, about ValuJet's safety problems before May 11, 1996, but failed to act.

The first suspicious event happened on April 23, 1996, almost three weeks before the crash, when Peña's own Department of Transportation released a report, "Low-Cost Airline Service Revolution." The report gleefully claimed, "In the last year, American consumers have saved an estimated $6.3 billion in airline fares" because of "new low-cost, low-fare airlines." Most of the credit, Peña said in a side statement, went to his boss, President Clinton.

Curiously, one of the airlines the report repeatedly referred to as "new" was Southwest Airlines, which had actually begun service in 1971, some fifteen years earlier. In fact, Southwest accounted for three-fourths of the flights of low-cost carriers and was by far the safest.

On May 2, 1996, nine days before the ValuJet crash, the FAA produced a nine-page report on the safety records of the various new airlines. The report was ordered by Anthony Broderick, the FAA's associate administrator of regulation and certification. Broderick, a man of integrity, resigned his position shortly after the ValuJet crash and is generally viewed as having taken the fall for Secretary Peña. Bob Matthews, an analyst with the FAA's Office of Accident Investigation, prepared the report. Matthews, another good man, provided two different sets of data—one including and one excluding Southwest Airlines. The results were startling. When Southwest was included, the overall accident rate of the low-cost carriers was virtually the same as that of the higher-priced carriers, and the serious accident rate was identical. But when Southwest, not a new airline at all, was excluded, the accident rate of the new low-cost airlines jumped to four times that of the high-priced carriers, and the serious accident rate was more than six times greater.

Virtually all the discrepancy between low-cost carriers and bigger, more established airlines was due to ValuJet. Far from "exceeding" FAA standards, ValuJet's record was a nightmare. While the other eight start-ups had one accident among them, ValuJet had five. The others had no serious accidents; ValuJet had three. ValuJet's accident record was fourteen times higher than that of the high-fare carriers. Its serious-accident rate was thirty-two times higher than that of the larger commercial carriers.

By February 7, 1996, the situation was so critical that staffers from the inspector general's office met with FAA inspectors in Atlanta to discuss ValuJet's problems. The role of the inspector general in the Department of Transportation is to independently assess the performance of the agency and to investigate possible wrongdoing. Documents show that the inspector general's staff addressed more than a dozen safety problems at ValuJet. Mary Schiavo, the inspector general of the DOT, said she decided right then, "I would not get on [ValuJet] because of the number of incidents that have been reported." An internal FAA memo distributed seven days after Schiavo's statement suggested that ValuJet be shut down pending recertification. On May 6, just five days before Flight 592 went down, an FAA preliminary draft report listed more than one hundred ValuJet safety violations.

Testifying before Congress on June 17, 1996, the day Secretary Peña finally ordered that ValuJet be shut down, Mary Schiavo said she had brought up the ValuJet problem repeatedly with Peña's chief of staff, Ann Bormolini. "To my amazement," Schiavo later told a reporter, "Peña and Hinson said they knew nothing about the ValuJet concerns."

It was a big, messy divorce, government-style. With the inspector general on one side and the FAA administrator and the secretary of transportation on the other, the FAA cracked wide open. David Hinson, the FAA administrator, was replaced later that same year. Peña resigned as secretary of transportation in 1997 but remained in the Clinton cabinet as secretary of energy. Mary Schiavo, the inspector general who couldn't be silenced, resigned. President Clinton then instituted a ninety-day safety review of the FAA and Senate probes into the aviation safety inspector's role in the ValuJet accident.

Before he left as FAA administrator, David Hinson appeared before a Senate committee investigating the FAA's aviation safety-inspector program. During his prepared speech, Hinson did not, even once, mention the lives lost on ValuJet Flight 592. It was Jessica Dubroff all over again, except multiplied by 110. Instead, Hinson criticized the Senate for failing to approve a budget request that would have allowed the FAA "to fulfill its vital functions, including the effective safety oversight of the air transportation industry," by hiring additional inspectors. David Hinson either was morally bankrupt or had balls of steel. He was, in effect, blaming the Senate committee for the ValuJet accident. He got the money too.

In fact, the FAA was given billions of dollars to initiate a top-to-bottom overhaul. The Certification and Surveillance and Evaluation Team (CSET), the Online Aviation Safety Inspection System (OASIS), the Airline Transportation Oversight System (ATOS), and many other programs were quickly established with the newly acquired funding. This action would have a great impact on my life and on my efforts to improve aviation safety when, in 1998, I went to work as the program manager for OASIS. There was also an immediate consequence for me due to the government largess, for Congress required meeting facilitators to be involved in all the programs. I was asked to train as a full-time facilitator, and although I did not know it, experience as a full-time facilitator would be my ticket to Washington in another two years.

CHAPTER TWENTY-FIVE

Burning Heaven

While investigators were still fishing pieces of the ValuJet plane out of the Florida Everglades, an even deadlier and more controversial crash occurred near New York City.

On the evening on July 17, 1996, TWA Flight 800 prepared to depart JFK International Airport in New York City for Paris, before continuing on to Rome. The scheduled departure was 7:00 PM, but the flight was delayed for an hour due to a disabled piece of ground equipment and a passenger-baggage mismatch. The aircraft finally pushed back from the gate about 8:02 PM.

The 212 passengers were the usual mix of Europe-bound flyers from JFK: an American composer, a French hockey player, the wife of a famous jazz musician, a French guitarist, an American film director, a crime-fighting advocate, a fashion photographer, plus college professors, executives, vacationers, and, on this night, sixteen members of a high school French club and their five chaperons.

After takeoff, TWA Flight 800 received a series of altitude assignments and heading changes in order to reach its cruising altitude. The last radio transmission occurred at 8:30 PM. A moment later, the cockpit voice recorder and the flight data recorder abruptly stopped recording, and Flight 800's radar transponder sent its last signal.

Within seconds, the airplane crew was notifying officials in New York and Boston about seeing "an explosion out here." Hundreds of witnesses on land and sea reported they had seen and/or heard explosions, accompanied by a large fireball, and observed debris, some of which were burning, falling into the water. About one-third of the witnesses reported seeing a streak of light moving upward in the sky to where the large fireball appeared. All 230 souls on board, over half of them Americans, were killed.

Even before investigators from the NTSB arrived at the scene the following day, initial speculation was that the crash was caused by a terrorist attack. Consequently, the FBI initiated a parallel investigation.

Two catastrophic disasters so close together made even those of us at the FAA Denver office question if flying was safe. Like many others, we also wondered if it was a terrorist attack. While terrorist attacks were not a worldwide concern in 1996, neither were they completely unknown or unexpected. The 1960s had seen a rash of airplane hijackings, mostly politically motivated.

On December 21, 1988, Pan Am Flight 103 was brought down by a terrorist bomb near Lockerbie, Scotland, resulting in 270 fatalities. Over half the fatalities were U.S. citizens; the attack was aimed at the "symbol" of the United States. Many of us recalled other terrorist attacks in the past, as well, including the 1972 Olympics in Germany and the 1993 World Trade Center bombing in New York City. At the 1996 summer Olympics in Atlanta, a bomb exploded, killing one person and injuring more than one hundred. Although September 11, 2001, was five years away, it was not hard to believe that a terrorist missile brought down TWA Flight 800.

Even my friend Norm Wiemeyer, manager of the Denver NTSB office and one of the lead NTSB investigators on site in New York, returned home convinced of a conspiracy by the U.S. government to hide the truth. The two most prevalent theories to explain the demise of TWA Flight 800 were that of a terrorist bomb on board and that of a missile striking the plane (attributed to an American military accident by some and to terrorists by others). Neither Norm nor any other NTSB official was allowed by the NSA and the FBI to interview credible firsthand witnesses who saw the explosion of TWA Flight 800.

The FBI claimed that allowing the NTSB to interview witnesses could interfere with a possible criminal prosecution. Even if the FBI claim made sense, public officials and agencies had lost their credibility. People in general, and those of us in the FAA in particular, had little reason to trust or believe whatever the government stated was the cause of the TWA Flight 800 tragedy, especially on the heels of the ValuJet fiasco.

Conspiracy theories make for great news, and around the world, people were glued to CNN as though it were Super Bowl Sunday or the World Cup. As I sat in the Denver office with a group of inspectors watching TV, we were oddly quiet. Each of us feared that the FAA would be blamed for Flight 800, as we were for Jessica Dubroff and ValuJet, and we secretly and silently felt that blaming terrorism was a reprieve for us. The reprieve lasted only until Alvin Dietrich* spoke up.

Alvin was an eccentric avionics inspector and a friend. Like my older brother, Clay, he had a genius IQ and a Columbo-esque approach to inquiries. He was the classic eccentric, stopping in the middle of a sentence to start work on an algorithm or showing up at work wearing mismatched shoes or, my personal favorite, with his face shaven on only one side.

"Wasn't a missile," Alvin said to us or to the television.

"Well, what do you think caused it?" I asked him.

"Ed Block, Ed Block." Other inspectors began rolling their eyes and fleeing the area immediately. As I said, he was eccentric.

"Who? What are you talking about, Al?"

"Ed Block's been saying for ten years that the Kapton and Poly-X wiring was gonna start bringing 'em down."

"That's more far-fetched than the missile story."

"No, it ain't. Using aromatic polyimide and aliphatic polyimide insulation on current-carrying wires is risky. Sooner or later, it'll crack and expose the metal wire to moisture and corrosion. Then the wires can arc and . . ."

"You don't really think a tiny, little spark caused a 747 to blow up," I said.

"No, Dave, but millions of little sparks did."

"If you say so, Al." I patted him on the back, turned off the television, and went back to my desk.

I once again sat spinning in my squeaky chair and wondering what really caused the crash of TWA Flight 800 and how it could have been prevented. Was it another example of faulty decision making, like Jessica and ValuJet, and if so, why did it happen? Why, why, why?

"Talking to yourself again?" Walt Wise asked.

I didn't realize that I had voiced my *whys* aloud. "May as well," I said. "Nobody else listens to me."

"Well, maybe other people know better than to listen to you." Walt threw a pile of Service Difficulty Reports (SDRs) on my desk. "Now that you know your limits, take a look at these SDRs for John Jamison*." Walt grinned and hurried off before I could come up with a retaliatory insult.

Jamison was an FAA supervisor who spent more time playing solitaire on his computer than he did actually supervising. This wasn't the first time I had been targeted when it was time to catch up on paperwork.

Service Difficulty Reports were sent by airlines to the Denver Flight Standards District Office (FSDO). They were required by FAR 121.703. This particular federal aviation regulation was written in 1964, when jet service was in its infancy; it contained a list of seventeen scenarios that must be reported to the FAA on the SDR forms. That was it. According to the FAA, there were seventeen things that would somehow magically tell the FAA whether or not an airline was safe, and those scenarios hadn't changed in thirty years. The Flight 800 scenario wasn't included in the magical list of seventeen scenarios. ValuJet's scenario wouldn't have required an SDR either. I groaned at the futility of processing the large pile of SDRs on my desk. It was no wonder airplanes fell out of the sky.

Then it occurred to me—what if Alvin were right? What if the cause behind the TWA Flight 800 crash were outdated and potentially dangerous wiring approved under some old regulation? What if the cause were faulty decision making?

I set aside Jamison's work and began studying the origin of the aviation regulations, looking for a pattern. Until 1978, aviation was a regulated industry. The FAA could require the airlines to implement critical safety improvements and then, because they also controlled ticket prices, could approve price increases to offset the cost of those safety improvements. This system assured safety was not weighed against profit and provided a mechanism for the FAA to quickly deploy critical safety improvements, with the costs shared equally by the entire industry.

After deregulation, the way in which the FAA approached safety improvements changed dramatically. The change was that the FAA had to provide proof that any proposed regulation would prevent future loss of life and that the benefit of the safety initiative outweighed the cost for both the government and the aviation industry. This was the same situation I faced at Air Methods when I refused to put wire-strike kits on helicopters. The proof of their value came *after* a disaster. The FAA can prove the safety value of a proposed change only by waiting for a disaster to occur, which proves its value. With deregulation, the FAA became a business stakeholder in the cost-benefit analysis. As a result, promoting aviation, a mandate, was easier for the FAA than improving safety, the other mandate. *Okay,* I thought, *this explains how ValuJet crept up on us, but what about TWA Flight 800?* To find an answer, I decided to dig through archives of aircraft accidents to find the first accident affected by deregulation.

That accident occurred on September 25, 1978, on a sunny California day, when PSA Flight 182 collided with a small Cessna. The Boeing 727 crashed into a San Diego neighborhood, killing all 135 people on board, the 2 men in the Cessna, and 7 people on the ground, including a family of 4. Nine others on the ground were injured, and twenty-two homes were destroyed or damaged.

The FAA responded to the disaster by attempting to accelerate the deployment of the Traffic Alert and Collision Avoidance System (TCAS), which had been engineered in the mid-1970s by the Mitre Corporation, a federally funded research-and-development center, and the Massachusetts Institute of Technology. This technology, which worked independently from ground-control systems, was capable of warning pilots of aircraft that may collide with them and recommending diversionary maneuvers. Prior to deregulation in 1978, the FAA could have required airlines to deploy TCAS immediately following the PSA Flight 182 collision and to pass the cost of deployment on to passengers. TCAS could have been deployed at a cost of about fifty cents per passenger. Had TCAS been deployed, it could have prevented hundreds of deaths, including the 135 passengers of PSA Flight 182, the 88 passengers of Ukraine Aeroflot in 1979, and the 67 passengers of Aeroméxico Flight 498 in 1986, a collision similar to PSA Flight 182. The latter ultimately spurred Congress and other regulatory bodies to take action, forcing the FAA to mandate TCAS collision-avoidance equipment.

What I found suggested that the FAA was not equipped to operate in a nonregulated environment back then, or even today. I believed I could show, given enough data, that the change from regulation to deregulation had turned the "friendly skies" into the "fiery skies." I would need data, lots of data, in order to draw a valid conclusion, and somebody would have to keep the data, to store it.

The Online Aviation Safety Inspection System (OASIS) was started in 1996. The ValuJet crash prompted Congress to throw billions of dollars at the FAA in hopes of fixing regulatory problems. OASIS was already being used to gather surveillance data, so I just needed to get them to include safety and operational data. I contacted Pauline Lowell, the information resource manager for the FAA, who was overseeing OASIS, to see how I could be more involved. Since I was an FAA inspector and facilitator, the best she could do was assign me to OASIS on a temporary detail. The data I collected through OASIS showed that deregulation made it harder for the FAA to implement safety improvements, and arguably, harder to protect the public. My analysis suggested that, in a regulated aviation world, the ValuJet accident wouldn't have happened and the TWA Flight 800 crash might have been prevented, unless, of course, it was a terrorist attack.

A year after the crash of TWA Flight 800, the FBI announced that no evidence of a criminal act had been found, so the NTSB assumed sole control of the investigation. The NTSB investigation ended with the adoption of its final report on August 23, 2000. The conclusion of the NTSB, as stated in the report, was that the probable cause of the accident was an explosion of the center-wing fuel tank, *most likely* a result of faulty wiring. This finding prompted me to search the SDRs to locate the number of reports of fires or other problems due to wiring.

I was amazed to learn that the FAA's seventeen safety scenarios did not require filing an SDR on a wiring problem, unless it led to a substantial fire or other substantial problems. I found only two SDR reports that even mentioned wiring. If wiring was not on the list of seventeen, it was not a problem. This was how the aviation world operated then—and now.

Of course, many people didn't believe the FBI statements or the NTSB conclusions. Some reports stated explosive residue was found on the back of some of the seats in TWA Flight 800. Although various explanations were given to explain its presence, this residue, along with inconsistencies and unanswered questions, continued to fuel conspiracy theories. Scores of books have been written about TWA Flight 800: some have praised FBI investigators, others have insisted there was a missile attack and a government cover-up, while still others have suggested the crash was caused by everybody doing business as usual, including the FAA's failure to do its job better. When I am asked what I think, I always say, "Ed Block, Ed Block," and look down at my shoes.

CHAPTER TWENTY-SIX

Pigs in the Sky

During all the disasters and subsequent upheavals in the summer of 1996, Nick Sabatini was promoted to director of Flight Standards for the FAA. I liked and respected Nick. He carried himself like a 1940s Italian mob boss—his very presence commanded respect. I had met him while I was learning the ropes as a new facilitator two years before his promotion. At the time, Nick was the FAA's Eastern Region manager. I was leading only my second work group the day Nick pulled me aside during a break. "You're really involved with the work group, Dave, and that's good," he said. I recognized the compliment as one that preceded a "however." I was right. "However," Nick said, "you need to be committed to the group, not just involved."

"I am committed," I barked in my defense.

Nick looked around the room, then pulled me out the door farther from the crowd before saying, "You and I had bacon and eggs for breakfast this morning, right?" I nodded yes. "That breakfast was provided by a chicken and a pig, right?"

"Uh-huh." It felt like a scene in *The Godfather*.

"Well, the chicken was involved by providing the eggs, but the pig, now the pig, the pig was not just involved, he was committed to providing us bacon." He looked me in the eye. "*Capiche*?"

I mumbled another "uh-huh." I got it. He was telling me that to improve aviation safety, I had to be willing to sacrifice everything and be truly committed.

Nick must have seen something in me that he liked, because he contacted me in 1996 and asked me to go to the Center for Management Training and Executive Leadership (CMEL) to become a national facilitator. It wasn't clear to me how becoming a full-time facilitator would improve aviation safety, but I said, "Okay, Nick. I'll bring the bacon."

I traveled a lot over the next two years, and my assignments as a national facilitator kept increasing in importance. It was not uncommon for me to meet with high-ranking government officials and corporate executives. Although my primary job now was as a facilitator, I didn't totally escape investigations, inspections, or paperwork. It was these old, familiar duties that got me in trouble.

In March 1998, FAA Supervisor John Jamison sent me to inspect Heli-Support Inc. (HSI) in Fort Collins, Colorado. Kevin Shields, the HSI manager, was a friend since before I worked for the FAA. Over the years, my job had required a professional distance, but I welcomed the visit.

The day I was there, while I was observing Kevin inspect an Aerospatiale Lama main rotor head, he said, "What do you know about criminal investigations?"

I watched him put a tag on the rotor head that declared it airworthy. As an FAA designated airworthiness representative (DAR), Kevin could determine the airworthiness of parts after repair and return them to service. "Criminal investigations?" I said. "Not much, I guess. Why?"

"Jeff Graves was here a while ago," Kevin said, "and he found the gate over there unlocked." He pointed to a locked chain-link enclosure. Since Heli-Support imported helicopter parts from all over the world, some with little or no service history, these parts had to be segregated from the main repair facility until approved for use.

"That's not good," I said. Jeff was a fellow FAA inspector who was assigned to HSI as well. "No parts ended up on a helicopter by accident, did they?"

Kevin insisted that none of the untested parts had been used by accident. It never occurred to me that Kevin would use an untested part on purpose. As I got ready to leave, Kevin said, "Jeff was really acting weird. You think I'm under investigation?"

"Just make sure your *i*'s are dotted and your *t*'s are crossed," I said and then got in my car and drove off.

Since I was assigned to perform surveillance on Heli-Support and wanted to know what was going on, I went to see Jeff. He told me that Kevin was being investigated and that some problems might be "criminal in nature." Jeff did not tell me any of the specifics. He simply told me to "keep it zipped." The manager of the Denver FAA office, Jeff Roy, subsequently requested that I stay away from Kevin and HSI until the investigation was complete.

It sounded simple, but the aviation world in Colorado and the surrounding western states is not large. When I attended a DAR training event in Las Vegas a few weeks later, Kevin Shields was there. Jeff Roy again advised me to stay away from Kevin. Considering our history and his presence everywhere I went, it wasn't easy to do.

After the Las Vegas event, Kevin called our office in Denver and asked if he was under investigation. Shortly thereafter, in August 1998, I was officially accused of having informed Kevin of an FAA investigation, which allowed Heli-Support "to hide or change evidence that may have shown they violated FAA regulations or the U.S. Criminal Code." I didn't see how anything I had said to Kevin could be construed that way, but everyone in the FAA office now saw me as a traitor. The more I thought about it all, the more furious I got. I had been judged guilty without evidence. Even if I had warned Kevin and, by doing so, had stopped him from using illegal parts for repairs and saved a helicopter from crashing and people from dying, would it have been wrong? Our primary concern should have been to take immediate action if Jeff had evidence that Kevin was truly allowing unsafe parts to be used. I couldn't understand why the FBI would wait for action when they suspected unsafe parts were being released by Heli-Support.

I wasn't merely angry. I was ready to quit and move, even though neither my wife nor my son would have been happy about that. Jill was in the early stages of a new career as an insurance adjuster, while Tyler was about to start high school. Fortunately for them, I had no job offer and nowhere to go. Instead, I called Pauline Lowell and increased my involvement with OASIS. These duties at least kept me somewhat separated from the others in the Denver FAA office.

To my knowledge, there were never criminal or civil charges brought against Kevin Shields or Heli-Support. In October, Special Agent Thomas Funke from the FAA Investigation and Internal Security Division reported that his investigation had "found no information indicating that [I] obstructed justice or engaged in any conflict of interest." I was cleared of any wrongdoing, but change was coming anyway.

In early 1999, Nick Sabatini became the associate administrator of safety for the FAA. This position placed him third on the organizational chart, so when Pauline contacted me, at Nick's request, to talk about OASIS, I listened.

The OASIS program was created to improve the job efficiency of FAA inspectors. Its main focus had been to provide laptops and customized software to the three thousand inspectors in the Flight Standards Division. OASIS also provided remote access to inspection procedures, forms, regulations, and other documents used during inspection, surveillance, and certification activities. It was all meant to allow us to spend more time in the field and less time in the office. Before OASIS, I carried two large cases of manuals and a satchel with steno pads and file folders. Even then, I rarely had everything I needed. With OASIS, I just needed my laptop and a portable printer.

While it was a good theory, OASIS, like most FAA projects, had failed in execution. I told Pauline that OASIS had lost hundreds of my surveillance records and other data and was severely flawed. Pauline stopped me. "David, that's why we want you. You're tech savvy and have field experience. We want you to be program manager."

Once again, I was amazed at how circumstances in my life had led me, unaware, to opportunities. Nick's offer would finally put me in a position where I could directly influence safety procedures, but I knew it would also put me in a high-exposure position. Did I really want to attend a congressional hearing to be grilled about OASIS? I wondered how I would fare in the world of political jockeying and institutional thinking. I had barely avoided being lynched by the FAA and the FBI. Instead of telling Pauline these concerns, I told her I needed to talk it over with my wife since it involved moving the family.

"But it doesn't," Pauline said. "You'll have to come to D.C. several times a month, but not move here."

I was running out of excuses. Since I knew Jill would be happy for me, especially if I didn't tell her how much travel would be involved, I accepted the job.

Within days of starting as program manager, Nick Sabatini called me into a board meeting with high-level FAA officials. They were debating whether to terminate the contract with Galaxy Scientific and scuttle the laptop program. Galaxy had already distributed over twenty-eight hundred laptops, portable printers, and peripherals to Flight Standards' offices, but the software programs were so faulty that inspectors were resistant to using them.

I needed time to review the matter. A week later, I advised Nick to continue the contract. Information exchange was necessary for good decision making, the key to safety, as I saw it. But due to the climate of secrecy in the aviation world, information exchange was a problem. OASIS wouldn't solve everything, but it was a step forward. Nick agreed but reminded me that OASIS was my program now, and I had to make it work.

I redesigned the software and removed the glitches, but with 150 FAA offices around the world, I had a big job ahead of me to sell everyone on OASIS. I had many nights alone in hotel rooms and time to ponder. *How can we structure data into information, information into knowledge, and knowledge into wisdom*? Day after day, I visited FAA offices, airlines, and repair stations, asking questions and listening to new ideas from everyone I encountered about how to improve safety. To assess and prioritize the importance of the ideas, I devised a way of measuring their usefulness. Does the idea improve the ability to detect threats and errors in flight? Does it provide safety critical information to pilots? Does it disseminate information to inspectors in the field to improve making critical safety decisions? Then I added financial and performance criteria to the assessment. Does the idea improve the efficiency and effectiveness of the inspector? Does it provide accountability through records of activity?

At each stop on the OASIS tour, I was learning more and more about management culture. I learned that FAA managers followed a firm set of rules. For example: unruly employee? See Step 5a, Employee Coaching. It reminded me of the FAA's use of the antiquated seventeen scenarios for identifying an unsafe aircraft. It was disturbing to see a similar checklist for humans.

In the end, it was not the problems with hardware or software that threatened OASIS the most. Instead, it was the management culture and institutional thinking. Our D-day came during a meeting organized by the Professional Airways Systems Specialists (PASS) union. I was telling inspectors about the virtues of taking their laptops with them when they went on surveillance activities, when a piercing voice in the rear said, "That is not approved!"

The voice belonged to Tina Neuman*. Tina had no aviation experience of which I was aware, but she knew the rules backward and forward. "There is no official procedure that allows for taking government equipment out of the FAA offices," she said. The fact that these procedures were written before the invention of laptop computers, or any computers, didn't matter. Following the rules was how Tina got things done, and I was breaking the rules. Tina not only insisted the laptops remain in the office, but they had to be locked to the desks with cables, as well. She, of course, quoted the rules to support her position.

Tina delayed the laptop program by rerouting money into various accounting and statistical-analysis program. This earned her a promotion. It also earned her an acronym derived from the first words I heard out of her mouth: TINA—That is not approved! Tina's attitude represented one of the difficulties in improving safety: the system rewards people who promote the system, not people who improve it! If the TINAs don't thwart you, the TOMs will. TOM—Totally obstinate male. The corridors of Washington literally reek of testosterone. Give a man a badge or control over the budget, and out pops a TOM. The director of FAA Flight Standards—I'll call him Tom—asked me to develop an OASIS feature to provide the Department of Defense with personal information on every U.S. certificated pilot, including medical and violation information. When I reminded him of our obligation to protect personal-health information in compliance with federal health record privacy laws, he became defensive: "It's my data; I can give it to whomever I want." He then asked another program manager to provide the data. These stories of TINAs and TOMs may seem somewhat trivial until you realize that every day these people are making decisions that could put your life and the lives of others at risk!

Despite TINAs and TOMs and many other obstacles, OASIS not only survived, but it eventually became a successful and reliable program. Today, the OASIS program is named Flight Standard Automation System (FSAS) and still serves FAA inspectors. Although I had a grand vision for OASIS, I spent most of my time fighting for its survival. OASIS was a line item on the capital-improvement expenditures on the president's FAA budget, and as program manager, I had to maintain the funding each year by showing that the program was contributing to making aviation safer. I believed it was, but hard proof was nebulous at best.

The first year I faced the funding circus, my job was made easier by the upcoming Y2K millennium. Everyone was afraid that at midnight on December 31, 1999, all computer systems would fail, and any airplane in the air would simply crash. If letting them believe this meant that I got funding for OASIS, I was willing to remain silent. By doing so, I not only got our budget approved, I even won an award. Maybe I would fare better in D.C. than I thought.

CHAPTER TWENTY-SEVEN

When the Sky Is Falling

As we neared the end of 1990s, the FAA talked and talked about how the information-technology system would fail and what to do about it. In the end, we did very little, and Y2K hysteria was just that. A few months into 2000, after receiving an award for ten years of service to the FAA (ten years, really?), I was also given an award from Secretary of Transportation Rodney Slater for "excellence in service to the U.S. Department of Transportation for Year 2000 preparedness." I tucked both awards under my arm, acknowledged the applause of my fellow FAA employees, and thought how ironic it was that the government so often rewards people for doing very little. In truth, the award was really an excuse for an agency celebration. Everyone at the FAA was relieved that we had survived the dawning of the millennium.

For that matter, everyone in the FAA was relieved just to see the old decade come to an end. The FAA during the 1990s had taken more punches than boxer Mike Tyson: the Alaska Airlines' cover-up, ValuJet, young Jessica Dubroff, TWA Flight 800, the *Los Angeles Times* article, plus a growing list of well-known people who had died in aviation disasters—musician Stevie Ray Vaughn, Senators John Heinz III and John Tower, rock promoter Bill Graham, Secretary of Commerce Ron Brown, Disney President Frank Wells, singer John Denver, the presidents of Rwanda and Burundi, astronauts, a governor, a boxer, race-car drivers, the first female fighter pilot, and on and on. The 1990s truly did end with a bang, but not because of Y2K. The *bang* was the sound of airplanes crashing. A year earlier, in July 1999, John F. Kennedy Jr. crashed his plane into the ocean, killing his wife, her sister, and himself. Three months later, professional golfer Payne Stewart and his entire crew died in their plane even before it crashed.

The Kennedy tragedy was the American version of Princess Diana's tragic death, but for me, the deaths of Stewart and his crew were more poignant because I knew what had happened. I knew that some mechanic in Florida had made the same dumb mistake I once made while at Gates-Combs Learjet—only this time, unlike when I messed up, the mistake was deadly.

On October 25, 1999, a Gates Learjet 35 left Florida and headed for Texas, where Stewart was to participate in a golf tournament. Five other people were on board. After air traffic controllers lost contact with the plane, Air Force fighter pilots intercepted it. Fearful that everyone on board was unconscious or dead, the military planes prepared to shoot down the Learjet to avoid a crash in a populated area. The news called it "the ghost flight."

As I listened to the descriptions of the frosted windshield and the lack of response in the cockpit, I said, "It's the safety valve. It's in backwards." I recalled how the backward safety valve nearly killed me and Ridgerunner years ago. I sat watching the news helplessly as Stewart's pilotless plane traveled fifteen hundred miles, finally running out of fuel and crashing in South Dakota. The impact made a hole forty feet wide and ten feet deep. The severity of the crash made it impossible for investigators to determine the true cause of the accident, but I knew. Sadly, I also knew that a simple ballpoint pen could have fixed the problem.

These days, I was no longer a mechanic toting a toolbox, an executive managing budgets, an investigator visiting crash sites, or a facilitator running meetings. Instead, my job was now to make OASIS the teller of truth. To do this, the software had to be modified to keep a permanent copy of every Program Tracking and Reporting System (PTRS) record, a history of what every FAA inspector did and did not do. Because our records were available through the Freedom of Information Act (FOIA) and because inspectors soon figured out they could use OASIS to inform the outside world about problems, OASIS also provided a new level of accountability for the FAA.

In fact, we designed the system so records could not be deleted by anyone except Regina Houston or her staff at the Volpe National Transportation System Center operated by the Department of Transportation. Regina was the division manager of Safety Information Systems. I knew of no one smart enough to outwit or bold enough to provoke Regina. One night, when I was telling my wife about Regina Houston, Jill said that I had found the first member of my team.

"Team? What team?" I asked.

"The team that'll help you save lives and prevent suffering, David," Jill said. "Isn't that what you're doing? Isn't that why we've sacrificed so much?"

Jill was right. I was not alone in my safety quest. For months after that conversation with Jill, I talked to Regina about how we needed to change the way we looked at accident prevention and how OASIS could facilitate this. She listened to my wild ideas, mated them to software-development methods, and implemented plans. Before long, OASIS contained Planning and Surveillance Modules, Critical Data Analysis Reports, Risk and Threat Probability Analysis, Surveillance Activities Based on Risks, plus other major developments.

Despite all the failures of the FAA as a whole, OASIS was working. For the first time in my aviation career, I felt I was making progress to improve aviation safety. I believed the OASIS program could stand up to just about any challenge, a belief that was soon put to a test when the most egregious collection of deadly mistakes made by the FAA brought down Alaska Airlines' Flight 261 on January 31, 2000. It crashed into the ocean off the coast of California, killing two pilots, three crew members, and eighty-three passengers. The prophetic words of Mary Rose Diefenderfer in 1993, warning the FAA of activity at Alaska Airlines that put the public at risk, activity that the FAA tried to cover up, had come true. But there was a difference.

When the NTSB investigated the crash of Flight 261, it did not blame Alaska Airlines but the FAA itself. The NTSB determined that lax surveillance and improper maintenance approvals by FAA inspectors, both of which were recorded by OASIS, were direct causal factors in the crash.

Alaska Airlines started as a regional airline serving the northernmost state, a place where roads are scarce and air travel short and frequent. Due to the frigid winter and the short flights, Alaska Airlines asked the FAA to exclude it from certain maintenance rules. The FAA agreed, approving an increase in the elapsed-time requirement for greasing the actuator parts from 350 hours of operation to as many as 7,500 hours. In addition, the time between inspections to determine the wear limits of these parts was extended. Lastly, the airline was allowed to change the grease used on the actuator to one better suited for extreme cold, but Alaska Airlines also used this thinner grease on the rest of its fleet, including planes flying warm-weather routes. These combined factors caused the horizontal-stabilizer actuator on Flight 261 to jam, sending the aircraft plummeting into the Pacific Ocean. The FAA decisions directly led to the crash and the eighty-eight deaths.

In response to public and official pressure following the disaster, Nick Sabatini initiated the commercial airplane Certification Process Study (CPS). The members of the CPS were to examine the certification, operations, and maintenance systems of the FAA. I was not part of the CPS findings team in 2000 and certainly had no idea I would become part of the CPS response team in another two years. For most of 2000 and 2001, I was too busy traveling and training inspectors on OASIS to think much about CPS or the future. But I was proud of the role OASIS had played in revealing the errors that led to the Alaska Airlines' tragedy.

By the time September 2001 rolled around, the awards from a year earlier were gathering dust, and I was once again returning to Denver from the East Coast after another round of budget-begging meetings to fund OASIS. The 2001 meeting was held at the Volpe Center in Boston. It was a brutal affair. I spent the weekend there and, anxious to leave, caught a red-eye flight that would get me to Denver around 6:00 AM on Tuesday, September 11. Jill was going to be surprised to see me home so early.

The Rocky Mountains broke through the clouds as the captain announced our approach into Denver. In the foothills, the grasslands were turning winter yellows and browns. The mountain peaks to the west of the city were already dusted with snow. As the plane touched down at Denver International Airport, I glimpsed the architectural marvel, the jutting tentlike structure on the roof of the terminal. In the early morning light, it looked like a mirage or a drawing found in a book of Middle Eastern fantasies. I was content to be home, happy with my success in Boston, happy with OASIS.

Not long after I arrived at DIA, my cell phone rang. I didn't answer at first because I was wrestling with my carry-on bag, even though the caller was my wife. But when Jill called back immediately, I figured it was urgent. "Hello." There was no answer. "Hello?" I said.

I heard a big sigh on the other end of the line. "Thank God you're okay." I was wondering why I wouldn't be okay and didn't say anything. "Where are you?"

"Here in Denver. I wanted to surprise you."

"Can you see a TV, David?" I looked back at the closest gate where I had seen a television. There was a crowd around it now, a big crowd that kept growing by the minute. On the screen, I saw plumes of smoke coming out of a tall building. I recognized it as one of the Twin Towers of the World Trade Center in New York City. "I was so afraid it was you," Jill said. "I thought it was your flight from Boston." Jill started to cry. I was waiting for her to stop before I asked her what she was talking about when I glanced at the television and saw an airplane slam into the second tower.

About this time, I first heard the words "terrorist attack" and finally understood what was happening. Next, I heard an announcement that the Denver airport was being shut down and sealed. "I'm getting outta here, Jill," I said. "They're going to lock us in. I'll call you later."

I got off the phone, ran to the nearest exit, swiped my access key, and bolted through the door into the train tunnel. Even the trains that carry passengers from one terminal to another had stopped running. Alongside the rails is an access walkway, so I plodded along it until I found an exit. I hurried through restricted passageways and secure doors and finally got to my car. What I did the morning of September 11, 2001, to get out of DIA was probably illegal. I didn't care. When the sky is falling, you don't worry about whose umbrella you pick up. You worry about saving your head.

I arrived at the FAA office to find everyone in shock. Alvin Sharp sat quietly at his desk, Walt Wise made no witty comments, and John Jamison stared at his solitaire screen without moving a card. I didn't even ask myself the question that I instinctively asked after every accident: how could it have been prevented? I sat in my chair and thought, without spinning or squeaking. Many people wept. Those who didn't fought to keep their tears at bay. I was anxious to hug my son and hold my wife and tell her I loved her. It was all I could think of to do.

On September 11, 2001, the FAA, surprisingly, rose to the occasion. It responded quickly to ground all flights into and out of the United States. The only airplanes flying were military. In the aftermath of the attack, the decisions made by the FAA and other agencies proved less meaningful. They were decisions designed to reassure the public that somebody was doing something about safety (to promote aviation is the main FAA mandate, remember?). As a result of security changes, our airports were turned into internment camps. Of course, we all knew that no color charts of threat levels or passengers half-undressing or any nonsense about three-ounce bottles of liquid being safe but four-ounce bottles not being safe would have prevented the 9/11 attack.

In the weeks following 9/11, I completely lost interest in OASIS. I stamped various reports "Approved" without even looking at them. In all the years I had anguished over how to improve aviation safety, I had never once considered the potential hazard of someone using an airplane as a purposeful weapon of destruction.

In what seemed now to have been the flicker of an eyelid, all the certainty about my life's direction and purpose had crashed around me as sadly and definitively as the Twin Towers had come tumbling down. When TWA Flight 800 was thought to have crashed due to a terrorist attack, I knew, at the time, that politics was the wild card in the aviation game. There were no lists, no regulations, and no safety programs that politics could not trump.

For two months, nothing could shake me out of my slump. My quest to improve aviation safety now felt as feeble as Don Quixote's battle with windmills. I considered quitting the FAA. I considered leaving aviation altogether. I considered giving up. Oddly, it was another airline crash that roused me. On November 12, two months after the terrorist attack, American Airlines Flight 587 crashed into a neighborhood in Belle Harbor, New York, adding another 265 victims to aviation deaths in 2001.

As I listened to the NTSB investigator describe the cause of the crash, I realized that terrorist attacks, since the beginning of aviation, had accounted for a minuscule number of deaths—less than 1 percent—compared to deaths as a result of mistakes made in boardrooms, maintenance hangars, and airplane cockpits. These mistakes I could still try to stop. These lives I could still try to save. God could sort out the rest.

CHAPTER TWENTY-EIGHT

Aviation Oversight

Once terrorism was ruled out as a cause of the crash of American Airlines Flight 587, the news media returned to showing more photos of the 9/11 terrorists. Like others who watched, I wondered how the FBI and the CIA missed knowing about them. After all, the terrorist pilots were U.S.-trained. I finally called an old friend in the FBI I'll call Bob. Bob explained that the FBI actually had two divisions, intelligence and enforcement, and the rules under which intelligence could share information about potential threats were strictly regulated. He described it as a "brick wall." If Bob was to be believed, and I did believe him, the 9/11 attacks succeeded, in part, due to poor information sharing that led to worse decision making.

This time, I didn't feel Newton's apple of insight plunk me on the head; I felt the tree shake. The same brick wall existed in aviation. In aviation, the wall separates airline from airline and the aviation business from the public. I couldn't believe I had overlooked this.

In December 2001, I explained our new mission to the OASIS team: to remove the barriers that prevent the sharing of information. "Nothing will improve safety more," I said. Regina Houston looked at me skeptically. She knew it meant changing the entire aviation culture.

While I laid out my vision, I saw John Allen enter the room and stand in the back. John was the deputy manager of the Commercial Certification and Surveillance Division and also a colonel in the National Guard. I could tell that John wanted to speak to me, so I quickly closed the meeting and followed him to his office. He invited me to sit. "Would you be interested in filling in for me?"

"Filling in?" I was confused. On the organizational chart, John was at least three levels above me.

"On the CPS response team," he said. "We're replacing the findings team, and I'm in charge of Information Sharing and Data Analysis. I need you to take my place."

The commercial-airplane Certification Process Study (CPS) response team included upper management from airlines, manufacturing, and the FAA, as well as experts in other fields. "I don't understand, John. Why?"

"I'm leaving for military duty and expect to be absent about a year." I started to speak, but he interrupted me. "I can't tell you anything more about it, Dave."

When I'm under stress, I make jokes, so I said, "Sounds like an offer I can't refuse."

John didn't laugh. "Good, 'cause I'm due at the Pentagon for a meeting. You'll get your reassignment papers on Monday." John wished me good luck, and we walked out the door together. The invasion of Iraq soon afterward answered all my unasked questions about John.

To my surprise, I was ready to leave the OASIS assignment. I had taken it as far as I could in improving the quality and availability of information for the FAA safety inspector. I needed to find another way to achieve my mission of reducing accidents and improving aviation safety, and the CPS response team was a good first step.

My new job title was national information technology and business process lead. I went to the Human Resources Office and received my transfer paperwork. As I started to leave, the woman at the front desk stopped me and said, "Don't forget your relocation orders."

"Relocation?" I nearly dropped the papers. When John offered me the job, he had said that I would need to move. Once again, I hadn't discussed the move with Jill, and I had conveniently forgotten that part already.

I flew home and spent hours extolling the virtues of Washington, D.C., and my new job. Jill interrupted my sales pitch. "We have to move there, don't we?" She thankfully had said "we," but I could see she was hurt. "David, next time, would you, for once in your life, ask me before you totally change my life?" I promised her I would, but I knew she didn't believe me. Why should she?

We sold our Denver house in December 2001 and rented a condo in Alexandria, Virginia, but Jill was seldom there. As the manager of a catastrophe-response company, she followed in the wake of hurricanes, hailstorms, and floods to settle insurance claims. Florida had suffered massive hurricane damage, and Jill was dispatched to go away on assignment. For a while it was romantic to be apart and then come together, but before long, it just felt lonely.

With Jill gone, I prepared for the first CPS response team meeting in Seattle. I was concerned about how I would do. The CPS findings team included the same people I saw as the cause of accidents—the decision makers in the aviation industry and regulatory agencies.

Nick Sabatini, the CPS board chairman, began the meeting by telling the fifty-plus people assembled in Seattle that we were going to review each of the fifteen negative safety findings the previous group had detailed. He referred to the CPS findings as "problems that needed to be solved." Our goal was to find solutions. To do this, we were divided into four teams and four subgroups.

As John's replacement, I was a co-leader of the Safety Information Awareness team, along with Holly Thorson from the FAA Certification Division and David Harrington from Airbus. Since nine of the fifteen findings were included in our team charter, it was considered an important group.

Our ten members were all highly accomplished professionals. Each person brought special skills into play, but the more we met, the more I realized that the vision of our team was limited by Nick's instructions. We were figuring out how to solve problems (the CPS safety findings) but doing nothing to stop the problems from occurring. As the months and meetings dragged on, and I heard the usual "change the regulation" and "make a list of fatal accidents," I challenged them to stop "bumper-sticker thinking." It won me a reputation as a "pain in the ass."

In 2002, we were invited to hold our CPS response team meeting at the Airbus manufacturing facility in Toulouse, France. Jill and Tyler went with me. We were able to tour France a little, and it was a fantastic experience to visit Paris as well as Châteauroux, where my two older sisters were born. My parents were sent there to France in 1956 after an Air Force officer failed to appreciate my father's practical jokes.

The trip was an important turning point for me. The opportunity to observe the European culture showed me that Americans and Europeans differ in more than just language and food. I could see that we also differ in how we think. Europeans seemed to have a more holistic approach, which was reflected in the way they manufactured their Airbus airplanes, while Americans seemed to be overly compartmentalized, with an assembly line mentality that was reflected at the Boeing manufacturing facility in Seattle.

As I watched an aircraft go through the assembly plant at the Airbus factory in Toulouse, I found that the workers in each separate phase were acutely aware of how their work affected the delivery schedule and the overall quality of the aircraft. The work teams had meetings throughout the week to share information about how processes could be improved and to discuss concerns about delivery schedules. Workers were routinely rewarded when they discovered ways to improve quality or to improve working conditions.

Conversely, the workers I had observed at the Boeing factory in Seattle, Washington, which our team had toured three months earlier, were encouraged to focus exclusively on the performance of their task. They were not concerned about how their work affected the delivery schedule or the overall quality of the aircraft. The Boeing work teams had very little communication with each other. Variance from procedures was met with criticism, and rewards were given to Boeing workers only when the job was performed perfectly according to procedure, without any variance.

I began to recognize that Boeing's culture reflected our American belief that there are systems in place that will protect us. But, I wondered, could an overreliance on safety systems, the systems designed to prevent accidents, actually be a problem rather than a solution?

The evening after touring the Airbus facility, I began to see that an overreliance on safety systems would lead to a higher probability of complacency in pilots, mechanics, and upper-level decision makers, whether airline executives or FAA management. I worked late that night at the small desk in my hotel room trying to find a way to explain this theory to the CPS response team the next day. At about 2:00 AM, from across the room, Jill threw a pillow at my head; it was time to turn off the light and come to bed.

The next day, I fought sleep as Chuck Huber, an FAA inspector and a member of another CPS response team, presented plans for revising FAR (Federal Aviation Regulations) 121.703 and its seventeen reportable events. As Chuck displayed a list of all commercial airplane accidents over the past ten years, he said, "From this list, we identified which critical systems have caused accidents more than once. If there was only one accident, we took the cause off the list."

Took it off the list? I couldn't believe it.

"What the hell are you thinking, Chuck? This is crazy! Why are we even here if we're not going to change things? Don't you think some team sat around like this fifty years ago discussing some magical list that would make aviation safe? After us, some other team will be discussing which new events to add to the thirty-seven critical systems you've come up with. And in the meantime, more people will die in plane crashes. We can't wait for accidents to happen. We need to predict and prevent them."

I finally took a breath and looked around the room. The faces I saw were either frozen in shock, red with anger, or blank from confusion. Only David Harrington of Airbus offered a smile and a nod. That was enough.

"The fifteen findings that CPS identified are just symptoms. The real disease is our reliance on procedures, rules, systems, and lists like these. We have to view safety differently." I saw another head nod, or maybe my rant was causing him a stroke. "We have to forget about chiseling the 'Ten Commandments of Crashes' in stone," I said, "and ask the right questions—questions that allow us to identify the threats and the precursors to an accident."

"And just how do we do that when the airlines won't share information?" a team member from the FAA asked.

An airline executive quickly responded, "We won't share information because the FAA will use it against us, and if they don't, the press will."

The room was suddenly alive with debate. We had finally hit the core issues. "You're like talking to a brick wall!" someone shouted. Then more people were shouting. As the group's co-leader and, more importantly, a trained meeting facilitator, I had to do something fast.

I took off my suit jacket and laid it over my chair. I then climbed onto the table and chirped out a quick whistle like a basketball game referee. The room immediately froze. "We need a new perspective," I said. "You, sir." I pointed to one man. "I can see a new side of you from up here. You have a bald spot. I hadn't noticed that." Some people laughed. The airline VP did not. I figured I would either change minds or be given a blindfold and a cigarette and be shot. "A new perspective—that's the key if we're going to do anything meaningful." I looked around the room again. All eyes were on me. "So what safety information are you willing to share?"

Mr. Bald Spot blurted out, "None," and then chuckled. "Well, I *could* tell you what parts fail on us before overhaul is even due, as long as you don't know who told you."

"That's good," I said. "So you need anonymity." I went around the room until everyone offered at least one piece of valuable safety information they were willing to share. Boeing would share reliability information if it could get more information about parts that were out of warranty, and Airbus would provide information about fuel efficiency in return for knowing more about the frequency and type of cockpit errors. It came down to this: if we made the identity of the source anonymous and removed the fear of retribution for reporting an error, everyone was willing to share information critical to aviation safety.

I worked all night examining notes, writing concepts, and drawing pictures and relationship maps. The next day, despite two sleep-deprived nights and a disgruntled wife, I stood in front of the CPS response team and presented the "Concept of Operations" proposal for what I called the Safety Information Sharing Environment (SISE).

SISE is a hardware a hardware/software platform that protects intellectual property and privacy. SISE would provide airlines, for the first time, with a way of openly and anonymously sharing safety information. The operational concept for SISE is simple: by combining and analyzing safety-related data and other pertinent information, SISE can be used to identify accident precursors, failure trends, and critical safety risks. Safety threats that could remain hidden in routine operations, resulting in a catastrophic aviation accident, now would be revealed. SISE is based on the same need that I identified while investigating the Aloha Airlines' incident: getting the right information to the right people at the right time.

Getting from concept to operations would take money, and lots of it. Like Oliver Twist, I held up my bowl to every research-and-development budget manager I could find and asked for more. Within weeks, I had a $500,000 budget to begin developing the proof-of-concept software. I also enlisted the help of Modulant Data Management Systems. Their tagline—"Bridging the gap between data strategy and execution"—was just what I needed. I spent two weeks at their office in South Carolina discussing each of the challenges that would have to be addressed in the development of the SISE software.

While the key element in getting airlines to openly share information was anonymity, anonymity was only the first hurdle. Sharing information about operations and maintenance wasn't all we needed from them. The airlines and manufacturers also had to share information about how they mitigated safety risks in their own organizations. This was of particular concern to them, since each airline spent millions of dollars engineering safety products and didn't want to give their results freely to competitors. SISE provided a way for them to share this intellectual property without sacrificing their legal right to own it. In fact, SISE provided a way for airlines to recover some of their development costs from other airlines by sharing.

Finally, it was important to find a way to verify that the information being shared was correct. Bad information could lead to disaster, so SISE also addressed this issue.

The process of selling SISE to the CPS board and to the FAA involved far more than presenting my initial concept paper in France. First, after receiving input from team members and others, I revised the concept for SISE. Second, I had to get the CPS response team behind SISE. I did, and the CPS response team eventually presented SISE to the Oversight Board of the Certification Process Study.

Nick Sabatini, who chaired the board, called a 7:00 AM meeting, the only time he had available. The director of Flight Standards, you may remember him as Tom, arrived grumpy after a long, early-morning commute, but others were excited to see a demonstration of what we had discovered.

I told the board that various airlines and aviation-industry manufacturers had volunteered to provide the information needed for testing SISE. I did not say whether the information they provided me was real or hypothetical, but I made sure to point out that none of this valuable information was required reporting under FAR 121.703.

"So what should I ask SISE?" I announced loudly to be sure everyone was awake. The room was silent. "Imagine you're a director of maintenance at an airline. What's the first safety hazard that comes to mind?"

I waited out a long silence before the vice president of safety at Northwest Airlines spoke up. "Main landing gear cracks on 737s. We had one fail and—"

"Hold it," I said, knowing that I had rudely interrupted him. "SISE doesn't discuss who, only what. Remember, you're like Batman. Real identity unknown."

Once we started looking at landing gear cracks, our four computer servers returned 117 "hits." Everyone was in disbelief that there were so many. As people shared more and more detailed information, I refined the search, looking for a common underlying factor to main landing-gear cracks. By the time we were looking at "the dirty dozen," the most severe incidents, we discovered that all of them were caused by the improper use of lock-tight fluid on a bolt that screwed into the housing. From these results, we speculated that mechanics didn't follow the maintenance manuals. Or did they? We needed to check the manuals.

I quickly linked into the maintenance manuals through the OASIS program and examined the procedures for the use of lock-tight fluid on the bolt. In the manuals, we found two contradictory instructions, one of which, if followed as written, would actually *cause* landing gear cracks.

My theoretical presentation was lauded, and I was complimented on the potential of SISE.

"Gentlemen, SISE shows more than potential," I said. "It shows that we need to send out a bulletin to have the entire 737 fleet inspected."

"Do you mean this is real data?" asked Jim Ballough, one of the CPS Board members. "I thought it was just a demonstration."

"It's both," I said. "It's real data from the airlines and a demonstration of how SISE can help us."

A silence fell over the room. For the first time, we had looked into the future and identified the next accident before it had happened. I knew then that SISE would work.

The CPS oversight board touted SISE as the number-one recommendation from of the CPS response team. In their recommendation, they said that SISE "could well provide the greatest impact on safety above any of the other safety recommendations." I hoped Mike Myers somehow heard them.

(See appendix for more information on SISE.)

CHAPTER TWENTY-NINE

Men Wearing Boys' Coats

FAA management assigned responsibility for safety-information sharing programs, including the recommendations from the CPS response team, to the Accident Investigation and Safety Analytical Services (ASIAS) division. In 2005, we provided ASIAS with the CPS response team final report, and our team was disbanded with little fanfare.

During the years I worked with the CPS response team, I also worked on the FAA's Systems Approach to Safety Oversight (SASO) team, initiated by the FAA to examine ways of improving the FAA's business processes and to examine the role of safety inspectors. Working with SASO felt like a return to OASIS, except that SASO was involved in all aspects of flight standards pertaining to software, information technology, and business-process development. In my position with SASO, I had two primary responsibilities: to help prepare the budget proposals for the FAA and Congress and to oversee new business-process-development safety projects. Working with SASO was going well when suddenly, in 2006, my career in the FAA came to a crossroad because of a call I received about SISE.

An airline executive who had worked with me on the CPS response team called to ask for my help. He was concerned that, rather than developing SISE as we intended, ASIAS had chosen to use NASA as a trustworthy third party to collect safety information from pilots, mechanics, airlines, and manufacturers and to ensure the information was not released to competitors or to the public.

NASA was collecting operations and maintenance information from pilots, mechanics, airlines, and manufacturers through phone and mail surveys. Those who provided information had no control over who had access to it. Others who had concerns about NASA's ability to protect the names of those it had sworn to protect withheld meaningful operations and maintenance information. I became concerned about ASIAS's ability to predict or prevent accidents if there was no meaningful information being collected.

I explained this problem to the director of ASIAS, Jay Pardee, who shared my concerns. He made several requests to FAA management for me to be temporarily assigned to ASIAS so I could help rectify the problem. Each time, the director of Flight Standards, again I'll call him Tom, rejected his request, claiming that the work I was doing with SASO was far more important. I later learned from an insider that Tom had rejected Jay's requests because ASIAS had already taken several Flight Standards employees from him, and he didn't want to lose another.

I became frustrated with the FAA's inability to get out of its own way. I told Jay I would continue to support ASIAS in any way I could, no matter what happened. I had no idea just how significant making that commitment would become.

Working with SASO, I generally spent two weeks in D.C., then two weeks at home in Denver. My supervisor, Joe Tintera, was the manager of AFS-600, the Regulatory Support Division at the Mike Monroney Aeronautical Center in Oklahoma City. So I also traveled to Oklahoma City about once a month. Despite my disappointment in the way ASIAS had deployed SISE, I was happy to be living in Denver and to be near Tyler, who had returned from college and lived in an apartment a few miles away. I had a family life again, and, as the end of the year approached, Jill and I looked forward to celebrating Christmas together at home. It was not to happen, at least not the way we thought it would.

Our festive plans were "scrooged" in early December when I received a letter from Debra Entricken. Joe Tintera was going to retire, and she would be my new supervisor in Oklahoma City. Debra didn't like it that I was usually late submitting my expenses and reporting my time and attendance. She also didn't like my free-spirited nature. In short, Debra was a TINA, and I was not her idea of a loyal and obedient FAA employee. The letter Debra sent said, "Report to work at the Oklahoma City Mike Monroney Aeronautical Center at 0800 January 6 or you will be considered AWOL." There had been no prior discussion, not even a warning. I would have to either sell my house and move in one month or be separated from my wife and son yet again. The last time Jill had been in Oklahoma, she ended up in the hospital due to an asthma attack triggered by the mold in the air. Her health was one reason Jill loved the dry Colorado climate.

I contacted Debra and explained my wife's health problems. I also argued that I could do my job better in Denver. I provided other reasons, but none fazed her. She said, "You either report as ordered, or you'll be AWOL, and you know what that means."

I knew exactly what it meant—it meant that Debra had me in her gun sight. Whether she was alone in the hunt or encouraged by others, I didn't know. The only thing I could do was go to Washington, D.C., to try to stop her or, failing to do that, find another position in the FAA.

On the flight to D.C., I settled into my seat and took out an article a friend had sent me called, "How to Survive an Airplane Accident." There were half a dozen, or so, tips. Some were Boy Scout phrases encouraging one to be a good citizen: "Listen to your safety briefing." Others encouraged one to choose the "safest" seat, within six rows of an exit, and also explained how to select the "safest" airline or aircraft. The writer could have included other tips—sitting in the back of the airplane, for instance. In fact, the list could have been many pages long. While I wouldn't call the safety tips useless, for under certain circumstances, some of them might help a person survive an airplane crash, there is no safety tip that would have mattered to the passengers of TWA Flight 800 or ValuJet Flight 592 or Alaska Airlines' Flight 261. No tips would have changed the outcome for the Promise Keepers or the little girl in Wyoming or Tic Tac Man.

I put the safety-tip article away and picked up another article provided by the same friend—"Top Ten Airline Safety Questions." One question asked, "Who decides what changes are made for safety?" The writer answered by saying, the "FAA and the civil air authorities." I laughed out loud. In almost every accident, the NTSB cites the FAA as one of the causes. A pilot flies into a mountain, and the NTSB says the FAA should have had a rule to prevent it. This is how and why rules get made.

My laughter caught the attention of the person beside me. "Airplane safety," I said in answer to his curious look. "What people don't know could kill them." He quickly looked back at his paperwork, and I continued reading, skimming through several inane questions about aircraft models and the type of emergencies most likely to occur. Then I came to: "How often do airlines crash?" Really, shouldn't that be question number one?

I put away the article and thought about my journey over nearly thirty years. So many times, it felt as if I were close to figuring out the single underlying cause of aviation disasters. So many times, I arrived at a solution, only to discover something else, something more.

In my own first attempt to evaluate risks and hazards, I taped a list to the lid of my red toolbox with a missing handle, broken off during the miraculous door-flap-landing flight with Little John. It read, "Fuel, weather, lightning, proper equipment," and so on. I added "Duty time and rest" to the list when Dave Hodges crashed after ninety days on flight duty. I included "Slow down and think" after Hooter and Bagel Boy perished on the flight I should have been on. Had I continued my list, it would have been much longer, but still useless. Events rarely, if ever, present themselves in the same way.

During my seven years as a young aviation executive, I came to believe that the willingness of airlines and manufacturers to sacrifice safety for profit was the root cause of accidents. There was a nonprofit version of this insight as well, as practiced by the FAA. The FAA routinely sacrificed aviation safety in favor of promoting aviation.

Then I switched careers, and this change brought about another shift in my views. During my seventeen years with the FAA, I came to believe that the underlying cause of aviation disasters was the inadequate gathering and sharing of information. If the right information gets to the right people at the right time, it greatly reduces the likelihood of a decision that would lead to a chain of events resulting in an airplane crash. This belief had led to the creation of SISE. But even while I was fighting for the implementation of SISE, it bothered me that people still made bad decisions after receiving good information.

To understand why this happened, I had reviewed many of the disasters I had investigated, as well as NTSB and FAA reports of other disasters. There is always a chain of failure events in an aviation disaster. I became convinced that it is always possible to trace the chain of failures back to the first link: somebody decided to let the helicopter pilot fly for ninety days straight; somebody decided to change the type of grease used on an airplane jackscrew; somebody decided to fly into a storm cloud; somebody decided to put explosive canisters in the cargo hold. Each faulty decision set off a chain of events that led to an aviation disaster.

But if faulty decision making is the underlying cause of most accidents, I then asked myself, are they really accidents? An accident is something that happens that we cannot control. Yet, I was convinced that we could reduce the frequency of aviation disasters. So how do I do this? What's next?

The answer wasn't long in coming. Debra Entricken refused to budge. She remained insistent that I move to Oklahoma City. Human resources claimed there were no other jobs open for me—a lie, as I later would learn.

The truth was that, over the past four years, while fighting for SISE, I had pushed, argued, cajoled, pleaded, and, at times, embarrassed people in the FAA. I had made enemies. I didn't take the FAA's crippling of SISE kindly or quietly, and as I was learning, the FAA didn't take my efforts to improve aviation safety kindly or quietly either. After talking it over with Jill (really, I did!), I decided to take an early retirement. The FAA called it "resign in lieu of involuntary separation." I called it being squeezed out.

As an added insult, my human resources officer informed me there would be no severance package. I dug through—and spent hours discussing with various sources—employee handbooks, FAA regulations, and Fair Labor Standard Act provisions. My work was rewarded when I found a clause requiring the FAA to pay a hefty severance settlement, a clause that the human resources director had overlooked. The money would provide me with income for over a year, but it did little to calm my anger and frustration. I turned in my inspector's badge, signed a stack of papers, and then walked through familiar hallways and corridors obtaining the signatures and checkmarks required by the "Separation of Service Checklist." No one I encountered so much as said good-bye. Finally, all the boxes were checked. After seventeen years, I was now officially "separated from service."

When I returned to the eighth floor, my human resources officer had already left, so I laid the checklist on her desk and looked across the vast rows of vacated cubicles. Each row sank into darkness as the janitor turned off the banks of fluorescent lights, one by one.

In the elevator, I hit the first-floor button with my fist, and the door slid closed. I watched the lighted numbers above the elevator doors descend, as if they were counting down to my departure—7, 6, 5, 4, 3. By the time I reached the second floor, I wanted to scream, but my clenched jaw wouldn't allow it. "First floor," said the familiar recording of a British woman's voice. Then the voice asked me to step out, and I did. It was the last time I would walk the halls of the FAA headquarters at 800 Independence Avenue.

It was a cold and dreary December night, but I still wandered about the Washington Mall for hours. I had no destination. That was the problem right there—I no longer had a destination. I wondered if I had managed to accomplish anything during my seventeen years at the FAA. Or even before. Had I improved aviation safety at all?

I kept walking, kept thinking. The Christmas lights hanging on 14th Street light posts swayed and chattered as the cold wind blew at my back. I pulled down the warm woolen hat I had brought with me from Colorado, folded my coat collar around my neck, and made a dash for the nearest cover, the Jefferson Memorial. I sat down on a frigid marble bench. Above me loomed a statue of Thomas Jefferson, and behind him, I noticed the words "with the change of circumstances, institutions must advance also to keep pace with the times."

How unfortunate, I thought, that the FAA didn't have a Thomas Jefferson to lead it. I looked up farther and read the beginning of Jefferson's words chiseled on the wall:

> I am not an advocate for frequent changes in laws and constitutions, but laws and institutions must go hand in hand with the progress of the human mind. As that becomes more developed, more enlightened, as new discoveries are made, new truths discovered and manners and opinions change, with the change of circumstances, institutions must advance also to keep pace with the times. We might as well require a man to wear still the coat which fitted him when a boy as civilized society to remain ever under the regimen of their barbarous ancestors.

I couldn't believe what I was reading. The words written nearly two hundred years earlier spoke to me as if they had been written that day. The words didn't merely inspire me—they freed me. I had left the FAA with a sense of failure. Now I suddenly saw my departure as an opportunity. Now I would work with the private sector to implement SISE. Now I would focus on going even beyond SISE and ASIAS. New technologies and complex processes require vision and daring. Now I was free of the confines of a system that resisted vision and daring, a system that I had outgrown.

Just then, a cab pulled up in front of the monument, and the driver honked to see if I needed a ride. I walked toward him with a new bounce in my stride. "Where to?" the driver asked.

"Shopping," I said. "I have to buy a new coat."

Epilogue

Since leaving the FAA in 2006, I have kept my commitment to Jay Pardee by contributing time as an industry representative on the ASIAS information-sharing work group, and more recently, on the Joint Planning and Development Office's Safety Management Systems (SMS) work group. My bookcase has become my new toolbox. I have read books that explore personal awareness and profound changes in people, organizations, and society; books on managing risk; books on crisis management; books on business structure and economic theory; even books that have an axe to grind with the aviation industry. I was accepted to the Harvard Executive Education Program to study strategic business planning, and I have enrolled in classes at the University of Colorado to study technical business management and performance-management engineering under Steve Ouellette and Dr. Jeffrey Luftig. When I wasn't reading or writing or giving talks on aviation safety, I was researching accidents on my own.

During this time, I have analyzed nearly every major air disaster. There are plenty from which to choose. From January 2007 to June 2010, there were 240 commercial-airplane accidents, 28 crashes on scheduled airlines that resulted in 1,795 deaths. In 2008, the FAA was accused of being in bed with Southwest Airlines, allowing a

Boeing 737 aircraft to fly with structural cracks in the fuselage, cracks that could have led to an Alaska Airlines–type disaster. In August 2009, a tourist helicopter collided with a small plane over the Hudson River, killing all nine passengers. In April 2011, Southwest Airlines Flight 821, a Boeing 737, made an emergency landing due to . . . you guessed it, a cracked fuselage.

In an article from *USA Today*, Bill Voss, the president of the Flight Safety Foundation, was quoted as saying, "If we continue at this pace, we'll be turning the clock back ten years on safety." Actually, Bill Voss is wrong. We're not turning back the clock; we're merely continuing to tick along as usual.

He may be correct with regard to the availability of meaningful operations and maintenance information. In January 2008, under pressure from Congress, NASA released the operations information it had obtained from pilots to the public, in violation of the agreement it had signed with the FAA to serve as the trustworthy third party and had sworn to protect. NASA's attorneys argued that releasing the data would potentially reveal the names of aviators the study had sworn to protect. NASA released the information anyway.

Of course, any form of travel—walking, running, biking, driving, boating, or flying—is, for the most part, more dangerous than never leaving the house. The national fatality rate while flying in an airplane is .03 deaths for every 100 million miles flown. The odds of being struck by lightning are greater than the odds of perishing in a plane crash. But it doesn't feel that way. Our fear of flying and crashing is far greater than our fear of being struck by lightning. It's not the statistical odds of aviation disasters that drove me all these years; it's the fact that most all the deaths could have been prevented. Yet, I have no doubt there will be more air disasters in the years to come.

To me, the only real solution to faulty decision making, the underlying cause of most aviation accidents, is to change how we think. Changing the way we think takes practice. I know that before the next aviation disaster, there will be a warning, a small change in a routine pattern, a threat that is a precursor of impending disaster. The warning will only be recognized if a pilot, an FAA inspector, an executive, a mechanic, a dispatcher, or even an airline passenger has changed the way he or she thinks and has practiced a sharpened mental awareness to recognize threats.

To aid in changing the way people think, I have developed a paradigm. The paradigm goes far beyond predicting an accident. The paradigm identifies the core capabilities required to make decisions—good decisions. Over time, I have defined and refined the paradigm until it has become REPAIR, which is an acronym for Recognize, Environmentalize, Prioritize, Analyze, Institutionalize, and Realize. It would take a separate book to detail how to use the paradigm and how it can be applied to aviation safety, health care, or business management; and in fact, that is my intent. In any case, I believe that REPAIR can restore our ability to recognize threats before they lead to disaster.

Even as we finished writing this book—a long, difficult, yet rewarding experience—and even as I was being asked by the Volpe National Transportation Systems Center if it could use this book for its safety classes, and even as I was being invited more and more frequently to speak about aviation safety, my intent has remained the same. During the past thirty years, my passion has been one thing and one thing only: to know that I have done my best to make your next flight safer. I know that by doing this, I am keeping my promise to Mike Myers. I know because I can hear Mike's voice saying, "Well done, Dave. Well done."

APPENDIX

History of the Federal Aviation Administration

EARLY DAYS

The Air Commerce Act of May 20, 1926, is the cornerstone of the federal government's regulation of civil aviation. This legislation was passed at the urging of the aviation industry, whose leaders believed the airplane could not reach its full commercial potential without federal action to improve and maintain safety standards. The act charged the Secretary of Commerce with fostering air commerce, issuing and enforcing air traffic rules, licensing pilots, certifying aircraft, establishing airways, and operating and maintaining aids to air navigation. A new aeronautics branch of the Department of Commerce assumed primary responsibility for oversight.

The Department of Commerce initially concentrated on such functions as safety regulations and the certification of pilots and aircraft. It also took over the building and operation of the nation's system of lighted airways, a task that had been begun by the Post Office

Department. In addition, the Department of Commerce improved aeronautical radio communications and introduced radio beacons as an effective aid to air navigation.

The aeronautics branch was renamed the Bureau of Air Commerce in 1934 to reflect its enhanced status within the department. As commercial flying increased, the bureau encouraged a group of airlines to establish the first three centers for providing air traffic control (ATC) along the airways. In 1936, the bureau itself took over the centers and began to expand the ATC system. The pioneer air traffic controllers used maps, blackboards, and mental calculations to ensure the safe separation of aircraft traveling along designated routes between cities.

The Civil Aeronautics Act of 1938 transferred the federal civil-aviation responsibilities from the Commerce Department to a new independent agency, the Civil Aeronautics Authority. The legislation also expanded the government's role by giving it the authority and the power to regulate airline fares and to determine the routes that air carriers would serve.

President Franklin D. Roosevelt split the authority into two agencies in 1940, the Civil Aeronautics Administration (CAA) and the Civil Aeronautics Board (CAB). CAA was responsible for ATC, airman and aircraft certification, safety enforcement, and airway development. CAB was entrusted with safety regulation, accident investigation, and economic regulation of the airlines. The CAA was part of the Department of Commerce. The CAB was an independent federal agency.

On the eve of America's entry into World War II, CAA began to extend its ATC responsibilities to takeoff and landing operations at airports. This expanded role eventually became permanent after the war. In 1946, meanwhile, Congress gave CAA the added task of administering the federal-aid airport program, the first peacetime program of financial assistance aimed exclusively at promoting development of the nation's civil airports.

The approaching era of jet travel and a series of midair collisions (most notably the 1956 Grand Canyon midair collision) prompted passage of the Federal Aviation Act of 1958. This legislation gave the CAA's functions to a new independent body, the Federal Aviation

Agency. The act transferred air-safety regulation from the CAB to the new FAA and also gave the FAA sole responsibility for a common civil-military system of air navigation and air traffic control. The FAA's first administrator, Elwood R. Quesada, was a former Air Force general.

FEDERAL AVIATION ADMINISTRATION

In 1967, a new U.S. Department of Transportation (DOT) combined major federal responsibilities for air and surface transport. The Federal Aviation Agency's name changed to the Federal Aviation Administration, and it became one of several agencies within the DOT. At the same time, a new National Transportation Safety Board took over the CAB's role of investigating aviation accidents.

The FAA gradually assumed additional functions. The hijacking epidemic of the 1960s had already brought the agency into the field of civil-aviation security. Since the hijackings on September 11, 2001, this responsibility is now primarily with the Department of Homeland Security. The FAA became more involved with the environmental aspects of aviation in 1968, when it received the power to set aircraft-noise standards. Legislation in 1970 gave the agency management of a new airport-aid program and certain added responsibilities for airport safety. During the 1960s and 1970s, the FAA also started to regulate high-altitude (over 500 feet) kite and balloon flying.

By the mid-1970s, the FAA had achieved a semi-automated air-traffic-control system using both radar and computer technology. This system required enhancement to keep pace with air traffic growth, especially after the Airline Deregulation Act of 1978 phased out the CAB's economic regulation of the airlines. A nationwide strike by the air-traffic-controllers' union in 1981 forced temporary flight restrictions but failed to shut down the airspace system. During the following year, the agency unveiled a new plan for further automating its air-traffic-control facilities, but progress proved disappointing. In 1994, the FAA shifted to a more step-by-step approach that has provided controllers with advanced equipment.

In 1979, Congress authorized the FAA to work with major commercial airports to define noise-pollution contours and investigate the feasibility of noise mitigation by residential-retrofit programs. Throughout the 1980s, these charters were implemented.

During the 1990s, satellite technology received increased emphasis in the FAA's development programs as a means to improvements in communications, navigation, and airspace management. In 1995, the agency assumed responsibility for safety oversight of commercial space transportation, a function begun eleven years before by an office within DOT headquarters.The FAA was responsible for the decision to ground flights on September 11, 2001.

The FAA ordered its inspectors on March 18, 2008, to reconfirm that airlines were complying with federal rules after revelations that Southwest Airlines flew dozens of aircraft without certain mandatory inspections.

MAJOR ROLES

The Federal Aviation Administration's major roles include the following:

– Regulating U.S. commercial space transportation

– Encouraging and developing civil aeronautics, including new aviation technology

– Regulating civil aviation to promote safety, especially through local offices called FSDOs (Flight Standards District Office)

– Developing and operating a system of air traffic control and navigation for both civil and military aircraft

– Researching and developing the National Airspace System and civil aeronautics

– Developing and carrying out programs to control aircraft noise and other environmental effects of civil aviation activities

FAA Mission

Our mission—Our mission is to provide the safest, most efficient aerospace system in the world.

Our vision—Our vision is to improve the safety and efficiency of aviation, while being responsive to our customers and accountable to the public.

Our values—Safety is our passion. We're world leaders in aerospace safety. Quality is our trademark. We serve our country, our customers, and each other. Integrity is our character. We do the right thing, even if no one is looking. People are our strength. We treat each other as we want to be treated.

List of FAA Administrators

Elwood R. Quesada (1958–1961)
Najeeb Halaby (1961–1965)
William F. McKee (1965–1968)
John H. Schaffer (1969–1973)
Alexander Butterfield (1973–1975)
John L. McLucas (1975–1977)
Langhorne M. Bond (1977–1981)
J. Lynn Helms (1981–1984)
Donald D. Engen (1984–1987)
T. Allan McArtor (1987–1989)
James B. Busey (1989–1991)
Thomas C. Richards (1992–1993)
David R. Hinson (1993–1996)
Jane Garvey (1997–2002)
Marion Blakey (2002–2007)
Robert A. Sturgell (Acting) (2007–2009)
J. Randolph Babbitt (2009–2011)
Michael Huerta (2011-Present)

The Nine FAA Responsibilities

FAA Order 8020.11 lists nine specific responsibilities for all accident investigations conducted by the FAA. The FAA investigations must determine whether or not the following were factors in the accident:

- The performance of FAA facilities or functions.
- The performance of non-FAA-owned-and-operated air-traffic-control (ATC) facilities or navigational aids.
- The airworthiness of FAA-certified aircraft.
- The competency of FAA-certified airmen, air agencies, or operators.
- The Code of Federal Regulations Title 14 Aeronautics and Space (CFR-14) was adequate.
- The airport certification safety standards or operations were involved.
- The operator/airport security standards or operations were involved.
- The airman medical qualifications were involved.
- There was a violation of CFR 14 Regulations.

My Analysis and Recommendations for FAA Changes Following the Review of the Aloha Airlines' Accident

Always ask these three questions:

- Was information available that could have prevented the accident?
- If information was available, did the right people have it at the right time?
- If the right people had the information at the right time, did they recognize it as critical safety information and take action to prevent the accident?

Determine if either was a factor in the accident:

- Critical safety information was available that could have prevented the accident but was not available to the decision makers prior to the accident.
- Critical safety information that could have prevented the accident was known prior to the accident but was not understood by decision makers as critical safety indicators or precursors of an accident.

Safety Information Sharing Environment Diagram

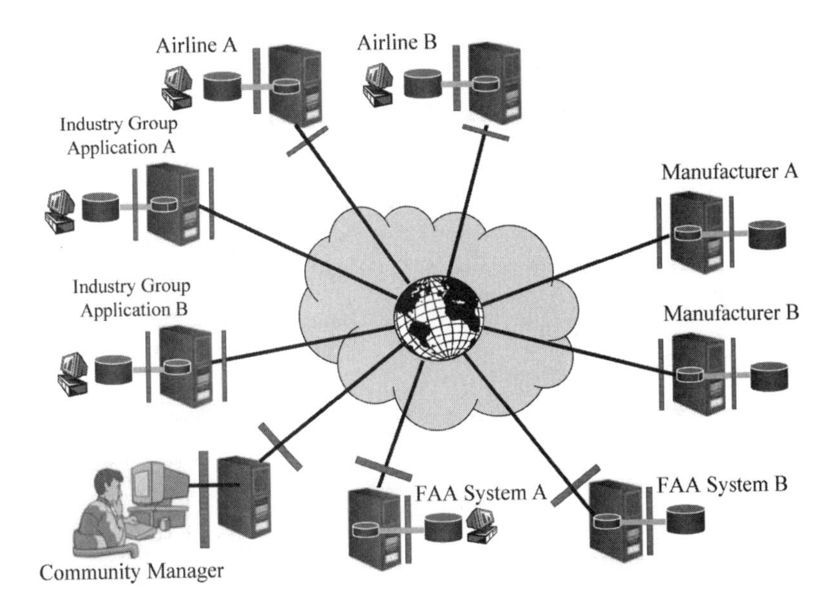

This diagram illustrates the concepts of SISE from the CPS Response Team report.

A member may choose to make most, or all, of its data available to the group and then specify controls over which individual members can access which data elements. For example, if Airline A wants to share a particular data element with Airline B and Manufacturer B but no other members, Airline A can specify controls that allow access to Airline B and Manufacturer B only. In SISE, there is no identification of who provides the data, and the provider is in charge of who is able to access it.

Contact David Soucie at dsoucie@whyplanescrash.com

Bibliography

The following is a list of works that includes information on the history of the Federal Aviation Administration and its predecessor agencies.

Bilstein, Roger E. *Flight in America*. Baltimore: Johns Hopkins University Press, 1984. 1900–1983.

Burkhardt, Robert. *The Federal Aviation Administration*. Frederick A. Praeger, 1967. -*CAB—The Civil Aeronautics Board*. Dulles International Airport, Virginia: Green Hills Publishing Co., 1974.

Davies, R. E. G. *Airlines of the United States Since 1914*. Washington, D.C.: Smithsonian Institution Press, 1972.

Gilbert, Glenn A., *Air Traffic Control*. Chicago: Ziff-Davis, 1945.

———. *Air Traffic Control: The Uncrowded Sky*. Washington, D.C. Smithsonian Institution Press, 1973.

Halaby, Najeeb E., *Crosswinds: An Airman's Memoir*. Garden City: Doubleday, 1978. An autobiography by FAA's second administrator.

Jackson, William E., ed. *The Federal Airway System.* Institute of Electrical and Electronics Engineers, 1970.

Kane, Robert M., and Allan D. Vose. *Air Transportation.* Dubuque, Iowa: Kendall/Hunt Publishing Company, 8th ed., 1982.

Kent, Richard J., *Safe, Separated, and Soaring: A History of Federal Civil Aviation Policy.* Washington: DOT/FAA, 1980. 1961–1972. See list of FAA historical publications in print for more information.

Komons, Nick A. *Bonfires to Beacons: Federal Civil Aviation Policy Under the Air Commerce Act.* Washington: DOT/FAA, 1978. 1926–1938. See list of FAA historical publications in print for more information.

———. *The Cutting Air Crash.* Washington: DOT/FAA, 1984. See list of FAA historical publications in print for more information.

———. *The Third Man: A History of the Airline Crew Complement Controversy.* Washington: DOT/FAA, 1987. 1947–1981. See list of FAA historical publications in print for more information.

Leary, William M., ed. *Aviation's Golden Age: Portraits from the 1920s and 1930s.* Iowa City: University of Iowa Press, 1989.

———. *Encyclopedia of American Business History and Biography: The Airline Industry.* New York: Bruccoli Clark Layman and Facts on File, 1992.

Pisano, Dominick. *To Fill the Skies with Pilots: The Civilian Pilot Training Program.* Urbana: University of Illinois Press, 1993. 1939–1949.

Preston, Edmund. *FAA Historical Chronology: Civil Aviation and the Federal Government.* Washington: DOT/FAA, 1998. 1926–1996. See list of FAA historical publications in print for more information.

———. *Troubled Passage: The Federal Aviation Administration During the Nixon-Ford Term.* Washington: DOT/FAA, 1987. 1973–1977. See list of FAA historical publications in print for more information.

Rochester, Stuart I. *Takeoff at Mid-Century: Federal Civil Aviation Policy in the Eisenhower Years.* Washington: DOT/FAA, 1976. 1953–1961. See list of FAA historical publications in print for more information.

Schmeckebier, Laurence F. *The Aeronautics Branch, Department of Commerce: Its History, Activities and Organization*. Washington: The Brookings Institution, 1930.

Thompson, Scott A. *Flight Check!: The Story of FAA Flight Inspection*. DOT/FAA, Office of Aviation System Standards, 1993.

Whitnah, Donald R. *Safer Airways: Federal Control of Aviation*. Iowa: Iowa State University Press, 1966. 1926–1966

Wilson, John R. M. *Turbulence Aloft: The Civil Aeronautics Administration Amid Wars and Rumors of Wars.* Washington: DOT/FAA, 1979. 1938–1953. See list of FAA historical publications in print for more information.